Introduction to Construction Health and Safety Management

First edition

Theo C. Haupt

juta

Introduction to Construction Health and Safety Management
First edition

Juta and Company (Pty) Ltd
First Floor, Sunclare Building, 21 Dreyer Street, Claremont 7708
PO Box 14373, Lansdowne 7779, Cape Town, South Africa

© 2022 Juta and Company (Pty) Ltd

ISBN 978 1 48513 291 2
eISBN 978 1 48513 292 9

Production specialist: Mmakasa Ramoshaba
Editor: Language Mechanics
Proofreader: Linda van de Vijver
Cover designer: Genevieve Simpson
Typesetter: Wouter Reinders
Indexer: Lexinfo

Typeset in Adobe Caslon Pro on 10.5/13 pt

This text is dedicated to all those construction workers who through their commitment to executing their allocated construction activities in challenging circumstances during their working lives have suffered the effects of poor construction health and safety management on the construction sites where they have worked.

It is also dedicated to Ferial, my unwavering support and partner, and our special four-legged children, especially little Maddie, our pride and joy.

FOREWORD

When the author, Theo Haupt, invited me to write the foreword to this fascinating book on construction health and safety, I agreed to do so for two compelling reasons.

First, having known and been involved in Professor Haupt's academic development for over twenty-five years, writing a foreword to one of his books is an act of great fulfilment and a privilege. Over time, I have witnessed and admired his passion for construction industry development generally, construction health and safety specifically, and love for South Africa, his beloved country in which context the book is written.

Second, he maintains a unique relationship with the construction industry; he might well say, with a nod to President Kennedy, 'ask not what your profession or industry can do for you; ask what you can do for your profession or industry'. Professor Haupt is a grassroots construction industry practitioner, activist and academic with a holistic perspective and a panoramic view of the interface between the construction process and health and safety in construction. Wearing this 'hard hat', he has challenged himself to make the construction industry a healthy and safer place in which to work and earn a living, and has become a leading global authority on health and safety in construction.

Construction activities can be hazardous when *occupational health and safety*, as is often the case, is taken for granted. The construction industry is one of the most dangerous industries in the world, where disproportionate incidents of death and injuries are reported frequently. While the financial and reputational costs to the industry are pronounced, the burden is disproportionately borne by employees and their families. This is particularly the case in developing countries, characterised by the absence of a social safety net, good health and safety education and regulations, and effective institutions. Therefore, the importance of a clear and simple health and safety construction manual, which this book is, cannot be overemphasised.

I have read many books written generally on *occupational health and safety* in construction and many written with regional and country specificities in mind. The novelty of this latest addition to the existing body of knowledge in health and safety is in its simplicity, arrangement of content, and the ability of the author to cater for all possible stakeholders in the construction industry. The book speaks *occupational health and safety* to clients, construction firms as entities, the disparate professions within the construction industry, regulatory authorities and, above all, construction workers, reminding everyone of their occupational health and safety obligations, irrespective of the size of the construction project.

The book encompasses various topical and contemporary construction occupational health and safety management issues in construction, ranging from the legislative and regulatory framework, risk analysis and hazard identification to managing occupational health and safety during uncertainty or the 'new-normal' brought about by COVID-19, among others. The book shows that effective occupational health and safety management in construction also makes a considerable contribution to the development of the industry in a variety of ways, including labour productivity and the image of the construction industry. My sincere hope is that the content, conclusions and recommendations will result in greater occupational health and safety practices in the South African, and by extension, the African construction industry generally.

Obas John Ebohon
Professor of Sustainability & Environmental Law
London South Bank University
March 2022

PREFACE

During the many years that I spent working on construction sites for various construction organisations of different sizes, I witnessed first-hand the harsh and ever-changing working environment and conditions that construction workers faced daily. Sadly, I also saw many workers injure themselves in the course of carrying out their construction activities. Driven by my desire to understand why and how construction workers would subject themselves to these daily exposures to hazardous situations, which if only simple interventions were introduced could be avoided, I decided to study at the University of Florida in the USA where I obtained a PhD. My doctoral dissertation examined a performance approach to construction health and safety given the dismal performance track record of this particular sector, which kills and maims a large cohort of people every year. I was inspired to conduct extensive research over many years by my interest in trying to understand the reasons for this poor performance and through that effort to contribute to its improvement. My research activities in construction health and safety have gained me both a national and international reputation of expertise in the construction health and safety space.

After publishing my earlier book, *Management of Safety, Health and Environment in South Africa: A Handbook,* which is a comprehensive text for use in general industry, I decided to write a series of books on construction health and safety to fill the gap in South African literature on the subject. This book, which consists of nine chapters, is the first in a series of three books and is designed specifically to introduce its readers to the management of construction health and safety at an elementary and foundational level. Other texts I have come across attempt ambitiously in a single volume to address the entire scope of the construction health and safety management space. The book draws on my own hands-on industry experience as well as the key findings of many completed relevant research projects on various aspects of construction health and safety. This experience is complemented by extracting from the broader existing body of knowledge on construction health and safety.

The book, by its very introductory nature, does not claim to cover every conceivable aspect of the subject. However, it provides foundational principles that can be used in almost any construction enterprise and on any construction site, irrespective of its size. The text provides introductory construction health and safety information required for practising at an entry or junior level within the construction health and safety professional space and as such will contribute to how these practitioners interact with the industry. The book will fill a gap in the existing literature by introducing current practice and procedures in the face of the new normal as they affect the construction industry in South Africa and elsewhere. It will provide the knowledge base underpinned by theories, methodologies and principles essential for the effective management of construction health and safety on construction projects both in the private and public sectors. If the principles espoused in this text are implemented, construction health and safety will be better managed within the context of the South African legislative and regulatory framework and better practice. However, the principles of construction health and safety management are universal and not confined only to South Africa. Therefore, the book should appeal to all construction health and safety practitioners everywhere. The focus of the text is not on mere compliance with legislation but on better practice, which should be the ultimate goal as part of a process of continuous performance improvement.

The book is an important companion for anyone wanting to know how better to manage construction health and safety. It provides guidance on key aspects and principles that can be applied on any construction site, irrespective of the size of the project. The book covers the occupational health and safety legislative and regulatory framework, the need for construction health and safety management, risk management, hazard identification and risk analysis, cannabis in the workplace, management of asbestos, and COVID-19, to mention a few topics. I hope that anyone using the book will be able to better manage construction health and safety through understanding the legal requirements for construction health and safety on construction sites; implementation of construction health and safety management systems; identification and mitigation of hazards related to construction activities; application of methods of hazard control and practical application of knowledge and understanding; and understanding the principles and procedures of effective construction risk assessment, hazard identification and risk mitigation.

Writing the book was challenging, given the dynamic nature of construction health and safety management, especially in the South African context, to ensure that the content was current and user friendly. To remain abreast of the latest developments I consulted many online resources for detailed information, some of which are listed at the end of each chapter as references for further reading. In some cases, the references are extensive while in others they are not. Each chapter has a series of simple review questions that are useful for training and evaluation purposes. The book can be read cover to cover or by referring to relevant sections in any order of preference.

Where I used online sources, I attempted to contact the sources for permission to include them in this text. In every case, I cited the source both in the text and in the reference list to ensure that the contribution was duly acknowledged and fully recognised, providing for readers to visit the sites themselves and familiarise themselves with the author/s first-hand. Where I received no response to my request for permission to use a graphic in the book, I included the website in the list of websites at the end of the book for readers to consult directly for more information.

I acknowledge that this text would not have been possible without part funding from the National Research Foundation (NRF) and that the opinions expressed and conclusions reached are mine and are not necessarily to be attributed to the NRF.

Prof Theo C Haupt
Durban, South Africa
March 2022

CONTENTS

CHAPTER 1
LEGAL PRINCIPLES

1.1 INTRODUCTION

When it comes to construction health and safety (H&S), the claim by a construction firm that it is complying with the law begs the question: Which law is it claiming to be complying with? A vast number of laws and regulations in South Africa and the rest of the world affect construction H&S either directly or indirectly. In many cases there are even overlapping requirements. The current construction H&S legislative and regulatory framework is complex and comprehensive. Despite the presence of these laws and regulations, the H&S performance of the construction sector continues to leave much to be desired as it remains consistently the sector that kills and maims the largest proportion of workers in the country. Furthermore, most H&S laws and regulations arguably merely set out the minimum requirements that must be complied with. Therefore, to make the claim of legal compliance suggests that the firm is aware, and satisfies the requirements, of every aspect of the H&S framework, which in reality is nonsensical, unrealistic and impractical. It also suggests that the firm is focused on only being willing to do the minimum with respect to H&S compliance, which has historically been shown not to improve the overall H&S performance of the firm and industry. The firm should rather be committing to and striving to do whatever it takes, including exceeding the legal requirements, to ensure the health and safety of every one of its workers as well as the public. A paradigm shift is required from an obsession with mere legal compliance with legislation, regulations, codes of practice, and minimum standards to better practice to improve the quality of life and the work environment of all workers and industry participants. However, taking cognisance of the focus of most construction firms to either make the minimum investment in construction H&S or not make any investment at all, laws are necessary to enforce minimum compliance.

1.2 THE NATURE OF LAW

It is universally accepted that laws guide society on what is right and wrong so that there can be peace and order in society. In the absence of laws, life would most likely be completely chaotic with everyone being a law unto themselves, with anarchy probably being the result. The law is typically defined as a rule of conduct or action prescribed or formally recognised in the body of rules as binding or enforced by a controlling authority, for example, a national parliament, provincial government, or metropolitan or local or district municipality. The law may also be viewed as the discipline and profession concerned with the customs, practices and rules of conduct of a community that are recognised by it as binding upon it. The law consists of 'legal rules' which deal with the conduct of individuals in society and apply to all members of the community. Therefore, it is important to know what the nature and practical effect of these legal rules are. Generally, it is accepted that all the members of a community will conduct themselves according to those customs and rules that they have accepted.

For instance, there are certain accepted linguistic rules that enable members of the community to communicate effectively with one another. For example, on construction sites workers use their own lingo or colloquialisms to communicate with each other about the various tools that they use and the activities that they execute on site. Someone who has not worked on construction sites before will not necessarily be able to understand what the workers are referring to.

Other rules are social ones, which enable members of the community to get along with one another. They would, for example, congratulate other people on their birthdays, and children are taught to always say 'please', 'thank you' and 'may I' when they want or receive something from someone else. If they do not, they could be considered rude or ill-mannered. Under COVID-19, the rules have changed somewhat. Instead, social distancing and avoidance of direct contact with others are practised as the new norm.

Legal rules are similar in nature, and regulate the rights and duties of each and every community member. On the other hand, non-legal rules apply to select groups in the community, for example, the rules of a sporting organisation or athletics club that apply only to the members of that organisation or club.

> *The main difference between legal rules and other rules of conduct in a community is that legal rules are binding upon members of the community.*

This claim means that community members may be forced to comply with these rules and that non-compliance with or ignoring them will have a certain prescribed detrimental effect on the person violating them. For example, should a person destroy property belonging to another member of the community, he or she may be punished by the state or ordered to pay compensation to the member of the community that has suffered damages as a result of his conduct. Going over the speed limit or driving without a valid driver's licence or driving under the influence of alcohol will incur consequences in the form of a fine and/or imprisonment and loss of driving privileges such as the confiscation of the driver's licence. Similarly, violation of the regulations governing the prevention and control of COVID-19 infections and prescribed protocols carries severe legal censures in the form of fines.

1.3 LEGAL TERMINOLOGY

To have a better grasp of the legislative and regulatory framework which governs the construction H&S industry in South Africa, the understanding of a few basic legal terms, concepts, doctrines and principles is useful. While the list of terms is not intended to be comprehensive or exhaustive, the following will be helpful.

Precedent
A prior or earlier ruling or judgment on any case in a court of law is known as a precedent and is a fundamental principle upheld by the courts.

Stare decisis/Precedent (to stand by decided matters)
This doctrine or policy of following rules or principles laid down in previous judicial decisions unless they contravene the ordinary principles of justice is a very important one, especially in the context of construction H&S. It is a legal doctrine that requires courts in South Africa and elsewhere to follow historical cases when making a ruling on a similar case with similar circumstances. *Stare decisis* ensures that cases with similar scenarios and facts are approached in the same way. Simply put, the doctrine of *stare decisis* binds courts to follow legal precedents set by previous court decisions. *Stare decisis* dictates that courts look to precedents when overseeing an ongoing case with similar circumstances. Lower courts are bound by the decisions of the higher court and are bound by decisions of the same divisional court, for example, KwaZulu-Natal or Gauteng. A recent prominent case involved former President Jacob Zuma, when he tried to overturn the order of the Constitutional Court by bringing an action in a lower court. The lower court ruled against Zuma based on the *stare decisis* doctrine and jurisdiction.

Appeal
An appeal is a request to a higher court to review a decision made by a lower court. The higher court can review the entire case, certain aspects of the case, or the sentence imposed by the lower court. The appellant, who may be the plaintiff or the defendant in the lower court case, must show the higher court that mistakes or errors were made during the previous trial, and that the case should be overturned, dismissed or re-tried. It is important to note that the right to appeal is not automatic but must be granted by the court handing down the judgment before the appeal process can commence. A recent example of this principle in practice was the famous Oscar Pistorius criminal trial when the prosecution, upon being granted leave to appeal the verdict handed down, appealed to the higher court and successfully had the sentence handed down made more severe based on the court's incorrect interpretation of the *dolus eventualis* principle.

Review
Review means to re-examine judicially or administratively; it involves a judicial reconsideration for purposes of correction of, for example, the examination of a case by

an appellate court. In recent history, reports and recommendations by the office of the Public Protector have been taken under review by, for example, the President and others allegedly implicated in those reports.

Locus standi

Every person whether natural or juristic has the right to bring an action, to be heard in court, or to address the court on a matter before it. Put another way, every person has the right to initiate proceedings in court and the right to be heard in court. The same principle is applied in terms of labour law in South Africa when disciplinary proceedings are instituted by an employer against an employee who allegedly contravenes a H&S rule in an organisation. The employee has the right to hear the charges made against him and the right to defend himself against those charges in a disciplinary hearing.

Liability

Liability refers to the legal responsibility for the acts or omissions of a natural or juristic person. Failure of a person or entity to meet that responsibility leaves him/her/it open to a lawsuit for any resulting damages or a court order to perform, such as in a breach of the terms of a contract or violation of a statute. The principle of liability is an important one when it comes to construction H&S.

Onus or burden of proof

Onus or burden of proof refers to the duty of presenting a certain amount of evidence by parties in order to meet the legal requirements for establishing the entitlement of one party in a case to the outcome sought. It is the obligation by parties to offer credible evidence in a court of law in support of a contention or accusation. In a civil matter the legal requirement is for the plaintiff to present sufficient evidence to prove its case based on the preponderance of probability, whereas in a criminal matter the legal requirement is for the State or prosecution to present sufficient evidence to prove its case beyond reasonable doubt. Failure to do so may result in the matter being dismissed.

Mandamus/Interdict (do something/stop doing something)

Mandamus or an interdict is an extraordinary writ or order that is issued by a court that commands or instructs a natural or juristic individual to perform, or refrain from performing, a particular act, the performance or omission of which is required by law as an obligation. It is made without the benefit of full judicial process, or before a case has concluded. It may be issued by a court at any time that is appropriate, but it is usually issued in a case that has already begun. An example would be the notice issued by an inspector from the Department of Labour to suspend operations on a construction site, which could be challenged by applying for an interdict to prevent the suspension from taking place.

Authorisations

Authorisations refer to the empowerment or empowering of another with the legal right to perform an action. Examples of authorisations include permits, licences and permissions. General power of attorney is a common form of authorisation giving someone the authority to act on behalf of another person as if they were that person.

Audi alteram partem (hear the other side (of the argument)/the right to hear and be heard)

Audi alteram partem has been quoted as being one of the most cherished and sacrosanct principles of law. This doctrine or principle embodies the concept that no person should be condemned, punished or have any property or legal right compromised by a court of law without having the court hear that person. It is the notion that every individual whose life, liberty or property is in legal jeopardy has the right to confront the evidence against him or her in a fair hearing. Therefore, no case or judgment can be decided without listening to the argument or account of another party.

Bona fide (in good faith)

Bona fide suggests that the person or party acts in good faith and is honest, genuine, actual, authentic, acting without the intention of defrauding. The law requires all persons in their transactions to act in good faith and a contract where the parties have not acted *bona fide* is void at the pleasure and discretion of the innocent or aggrieved party.

Mala fide (in bad faith)

A person acts in bad faith through the fraudulent deception of another person; the intentional or malicious refusal to perform some duty or contractual obligation. A person acts in bad faith by committing an intentionally dishonest act by not fulfilling legal or contractual obligations, misleading another, entering into an agreement without the intention or means to fulfil it, or violating basic standards of honesty in dealing with others.

Delict

In civil law, a delict is an intentional or negligent act that gives rise to a legal obligation between parties even though there has been no contract between them. It is the act by which one person, by fraud or malignity, causes some damage to some other. In its broadest sense, this term includes all kinds of crimes and misdemeanours, and even the injury that has been caused by another, either voluntarily or accidentally without evil intention. Delicts may be either public or private. The public are those that affect the whole community by their hurtful consequences while the private is that which is directly injurious to a private individual. Because of a delict's civil nature, the plaintiff must prove on preponderance of probabilities that they have suffered damage. So, in a civil action, if the plaintiff proves that it is more likely than not that the defendant was responsible for the injuries caused or loss suffered, the plaintiff wins.

The following are the key elements of a delict:
- Conduct: The plaintiff must prove that there was a voluntary/human/positive act or omission by the defendant that caused harm or damage to the plaintiff
- Wrongfulness or unlawfulness
- Culpability, which refers to the element of blame
- Causation
- Damage or harm suffered by the aggrieved party

The essence of these elements is captured in section 86 of the Mine Health and Safety (MHS) Act 29 of 1996 as amended:

Negligent act or omission

(1) Any person who, by a negligent act or by a negligent omission, causes serious injury or serious illness to a person at a mine, commits an offence.

Similarly, it is captured in section 37 of the Occupational Health and Safety (OHS) Act 85 of 1993 as amended:

(1) Whenever an employee does or omits to do any act which it would be an offence in terms of this Act for the employer of such employee or a user to do or omit to do, then, unless it is proved that –

(a) in doing or omitting to do that act the employee was acting without the connivance or permission of the employer or any such user;

(b) it was not under any condition or in any, circumstance within the scope of the authority of the employee to do or omit to do an act, whether lawful or unlawful, of the character of the act or omission charged; and

(c) all reasonable steps were taken by the employer or any such user to the employer or any such user himself shall be presumed to have done or omitted to do that act; and shall be liable to be convicted and sentenced in respect thereof; and the fact that he issued instructions forbidding any act or omission of the kind in question shall not, in itself, be accepted as sufficient proof that he took all reasonable steps to prevent the act or omission.

1.4 COMMON LAW

Common law is described as a body of unwritten laws based on legal precedents established by the courts. Common law influences the decision-making process in unusual cases where the outcome cannot be determined based on existing statutes or written rules of law. Common law, also known as case law or precedent, is the body of law that has been developed over time by judges and magistrates through court decisions rather than through legislative statutes or the normal course of proclaiming laws. Common law relies on detailed records of similar situations and statutes because there is no official legal code that can apply to a case at hand. Common law is based on the principle that it is unfair to treat similar facts differently on different occasions;

common law therefore binds future decisions where the facts and circumstances are similar. For example, if one person has been given a life sentence for rape, everyone else who commits rape should receive the same sentence.

> *Common law is often referred to as 'unwritten law' because it cannot be found in our legislation.*

Common law consists of those legal rules that have been in existence for generations, and that have through the ages been recognised by people as being binding upon them. Those legal rules are undocumented and unwritten.

Legal cases involving common law are interpreted in the light of the common law tradition. Therefore, they leave a number of things unsaid because they are already understood from the point of view of existing case law and custom. Put another way, what happens now is viewed and based on court cases concluded in the past. The common law applicable in different countries differs to a certain extent from country to country.

South African common law is known as Roman-Dutch law, because it is that which applied in Rome during the time of the Roman Empire. It had been refined by the Romans before it eventually became applicable in the Netherlands and, more particularly, the Province of Holland. It had developed further in Holland before being introduced at the Cape by Jan van Riebeeck around 1652, and then being made applicable in South Africa.

Despite the fact that much of South African contemporary law consists of legislation, important parts of it are still not regulated and are still part of South African common law. Most well-known crimes such as murder, culpable homicide, assault, rape, robbery, theft, housebreaking and extortion are not defined in legislation and are still part of South African common law.

In order to determine what the common law provides with regard to a specific matter, one has to consult the reported judgments of the courts in which the courts explain what the common law provides with respect to that particular matter. In order to determine which reported judgments to consult in a particular instance, one may consult the textbooks by lawyers dealing with that particular topic or case law records.

1.4.1 BASIC PRINCIPLES OF COMMON LAW

Statutes, which are discussed below, are understood to always be interpreted in the light of the common law tradition, and so may leave a number of things unsaid because they are already understood from the point of view of pre-existing case law and custom. By contrast to the statutory codifications of common law, some laws are purely statutory and may create a new course of action beyond the common law. An example is wrongful death, which allows certain persons, usually a spouse, child or estate, to sue for damages on behalf of the deceased. Where the claim is grounded in common law, all damages traditionally recognised for that delict may be sued for, whether or not there is mention of those damages in the current statutory law. For example, a person such as a breadwinner who sustains bodily injury through the negligence of another may sue for medical costs, pain, suffering, loss of earnings or earning capacity, mental and/or emotional distress, loss of quality of life, disfigurement, and more. These damages need not be set out in statute as they already exist in the tradition of common law.

Common law is based on the 'reasonable person' principle. It is not easy to define a 'reasonable person' and make it applicable in general. In fact, it is not possible at all to make such a definition. Each set of circumstances will determine its own 'reasonable person': in other words, what a reasonable person would have done under the specific circumstances.

In the context of construction, the following questions need to be answered if the worker is to be considered 'reasonable':
• Could the worker in question have foreseen that the incident or accident would occur on the construction site or workplace?
• Would the worker who had foreseen the incident or accident have implemented the necessary corrective measures?

1.5 STATUTORY LAW

Statutory law or statute law is written law passed by a body of legislature as opposed to oral or customary law; or regulatory law promulgated by the executive or common law of the judiciary. Statutes may originate with national or provincial legislatures or metropolitan, local and district municipalities. In those instances where common law did not provide for situations that had arisen in modern society, the State stepped in and promulgated statutes to deal with these situations. By far the majority of crimes in South African criminal law are statutory crimes.

Whenever one wishes to establish exactly what the law provides on a specific topic, it is always safest to ascertain whether there is any statute which deals with that topic or not. If no such statute exists, one would then have to turn to common law for a solution to the problem.

1.5.1 SOURCES OF STATUTORY LAW

The sources of statutory law in South Africa include the following:
• Statutes adopted by the Parliament of the Republic (Acts of Parliament) as the supreme legislative authority
• Statutes of the former republics, homelands or bantustans and colonies before they were included in either the democratic Republic of South Africa or the Union of South Africa in so far as they had not been repealed or amended

- Ordinances, proclamations and regulations of the various provincial authorities
- City councils, municipalities, divisional councils and village management boards, which are subordinate legislative bodies authorised to pass by-laws and regulations; by-laws and regulations such as traffic regulations are adopted with the consent of the Provincial Government or Parliament of the particular province
- Proclamations and regulations that have been promulgated by the President, Ministers and Premiers of provinces.

> *The purpose of statutory law is to satisfy the changing needs and demands of the country. Statutory laws are Acts, ordinances and by-laws introduced by bodies with legislation powers.*

1.6 THE SUPREME LAW IN SOUTH AFRICA

The constitution of a country contains those sets of laws that establish a state; an array of laws that *constitutes* the state, in the sense that the state is established, exists, and operates within the parameters of those rules. The birthplace of the South African construction H&S legislative and regulatory framework is the Constitution of the Republic of South Africa Act, 1996, as amended. The Constitution entrenches and protects the basic human rights of all people in South Africa, which includes all workers.

The Constitution, as amended, is the supreme law in South Africa. Therefore, any law or conduct inconsistent with it is invalid and the obligation imposed by it must be fulfilled. It supersedes all other rules contained in statutes, common law and custom. Furthermore, any rule inconsistent with a constitutional rule is an invalid rule. Any conduct that contradicts the Constitution, including failing to fulfil an obligation imposed by the Constitution, is also invalid. No Act of Parliament or regulation or municipal by-law can contradict the Constitution, and if it does, it can be challenged in the highest court of law in South Africa, namely the Constitutional Court. A recent prominent example of the supremacy of the Constitutional Court involved former President Jacob Zuma, and his unwillingness to appear before the Zondo Commission of Inquiry. The Constitutional Court ordered him to be incarcerated for a period of 15 months for failing to comply.

Chapter 2 of the Constitution, which is commonly referred to as the Bill of Rights, guarantees certain rights to various persons in South Africa, which include the right to an environment that is not detrimental to the health or well-being of individuals. The guaranteed rights include the following topics/elements:
- Equality
- Human dignity
- Life

- Freedom and security of person
- Privacy
- Freedom of religion
- Freedom of expression
- Freedom of association
- Residence
- Property
- Environment
- Housing
- Health care
- Children
- Education
- Culture
- Access to information

The Constitution entitles private parties to institute action to enforce their environmental rights against any other private party or corporation. A law or conduct can be unconstitutional because it violates a right in the Bill of Rights.

Section 24 of the Constitution of the Republic of South Africa, 1996, as amended, provides as follows:

> Everyone has the right –
> (a) to an environment that is not harmful to their health or well-being; and
> (b) to have the environment protected, for the benefit of present and future generations, through reasonable legislative and other measures that –
> (i) prevent pollution and ecological degradation;
> (ii) promote conservation; and
> (iii) secure ecologically sustainable development and use of natural resources while promoting justifiable economic and social development.

Therefore, in the absence of any construction H&S legislative and regulatory framework in South Africa, the provisions in the Constitution would provide such a framework. The question to be answered is: What if there were no construction H&S laws and regulations? How would employers behave toward their workers?

South African law is divided into different sections. Criminal law, the law of criminal procedure and the law of evidence are examples of such sections of the law.

1.7 CRIMINAL LAW

Criminal law is a body of rules and statutes that defines conduct prohibited by the State because it threatens and harms public safety and welfare and that establishes punishment to be imposed for the commission of such acts. In particular, it defines criminal offences, regulates the apprehension, charging and trial of suspected persons, and fixes penalties and modes of treatment applicable to convicted offenders. Conduct which is punishable by the State is known as a crime. Criminal law therefore deals with crimes and the punishment of the guilty parties.

1.8 REGULATIONS

An Act of Parliament tells us WHAT to do and not to do in terms of construction health and safety matters. The details of HOW to do and not to do are prescribed in various regulations, particularly the Construction Regulations, which form part of the Act.

Regulations are made by the Minister of Labour and form part of the related Act. The role of regulations in the construction industry is to prescribe standards and procedures that address construction-related health and safety issues unique to the construction sector. Examples of the wide range of regulations pertaining to the OHS Act include:
*   Asbestos Regulations
*   Construction Regulations
*   Electrical Installation Regulations
*   Electrical Machinery Regulations
*   Environmental Regulation for Workplaces
*   Explosives Regulations
*   Facilities Regulations
*   General Administrative Regulations
*   Hazardous Biological Agents
*   Hazardous Chemical Substances
*   Lift Escalator and Passenger Conveyor Regulations
*   Noise Induced Hearing Loss

1.9 CODES OF PRACTICE

The regulations and related codes of practice are an excellent guide to identifying and controlling hazards in construction. A comprehensive knowledge of those relevant to the construction operations and activities of an organisation is imperative if exposure to hazards on construction projects and sites is to be effectively managed.

Importantly, codes of practice are not legislation. Codes provide advice on how to comply with specific parts of the various regulations. For example, recently, the new Ergonomics Regulations, promulgated in December 2019, are another element to be incorporated into the already existing H&S programmes of the employer. They aim to eliminate, or reduce, ergonomic hazards in the early stages of designing systems and equipment, rather than trying to find a solution once there is a final product. Codes of practice, while not laws in themselves, need to be developed taking cognisance of the provisions and requirements of these regulations.

The construction industry is one that has many potential hazards to H&S, which could result in accidents and occupational diseases, the consequences of which could be that workers and the public are injured, disabled or even killed. It has been asserted that no matter how aware one is of the H&S factors related to work-related tasks and activities, the incidents or accidents will never be eliminated completely. This assertion needs to be challenged because it is based on a false premise that incidents or accidents are inherently part of the job. Furthermore, human factors as well as environmental factors play a major contributory role in the causes of accidents. It has been claimed that human beings are by nature lazy and will always look for a shortcut to complete a task or assignment and will often 'get away with it', but then will run out of luck, and someone will get hurt or killed. This claim raises many questions and should also be challenged.

It has also been asserted that ...

accidents don't just happen –
they are caused by unsafe acts
and/or unsafe conditions.

The reality, in fact, based on multiple research and analytical studies, is that all accidents in construction are failures of management and the management system and are therefore preventable.

Figure 1.1: A safety sign on a construction site in India

1.10 STANDARDS

A 'standard' refers to any provision occurring in a specification, compulsory specification, code of practice or standard method as defined in section 1 of the Standards Act 29 of 1993.

> *Standards can also be defined as the 'minimum requirement' against which performance is measured.*

The South African Bureau of Standards (SABS) is a South African statutory body that was established in terms of the Standards Act 24 of 1945, and continues to operate in terms of the latest edition of the Standards Act 29 of 2008 as the national institution for the promotion and maintenance of standardisation and quality in connection with commodities and the rendering of services. In the past the SABS, as the national standardisation authority, was responsible for maintaining South Africa's database of more than 6 500 national standards, as well as developing new standards and revising, amending or withdrawing existing standards as required. In recent years, however, the role of the SABS has changed. Specifications are now being done by SANS – a separate, independent body that reports to the Department of Trade and Industry.

The SABS is a certification body that is accredited by the South African National Specifications (SANS). It is therefore incorrect to refer to the SABS as a standard, as it is a testing and certification body that is allowed to sample and test products and certify a producer's product to a specific SANS standard, through their SANS accreditation. SANS, on the other hand, refers to a standard that specifies the performance requirements of a specific product. SANS standards are not the property of SABS. Most of the specifications are done by SANS and have either been developed nationally, or are standards adopted by the International Organization for Standardization (ISO). The thinking behind this is to bring the SANS specifications in line with international specifications as far as possible. Some confusion might still exist with the general public, seeing that SANS specifications are still obtainable from the SABS in Pretoria and SANS uses the same building as the SABS.

Most of the standards used in industry in South Africa were originally set by the SABS. For example, 'SABS 0177: Part II' means the South African Bureau of Standards' Code of Practice entitled Fire Resistance Test for Building Elements, SABS 0177: Part II – 1981. This code of practice will dictate the minimum requirements in terms of fire resistance to which building elements must conform.

> *These standards become part of regulations, which in their turn form part of the Standards Act.*
> *The Occupational Health and Safety Act 85 of 1993 governs health and safety in the workplace.*

1.11 BREAKING THE LAW

1.11.1 LEGAL LIABILITY

In various instances employers rely on insurance against liability for damages. This cannot be done for all liabilities faced by the employer, especially in construction. The materialisation of the risk faced could lead to either a civil or criminal action instituted against the employer or workers themselves.

In South African law, three basic forms of liability exist, namely civil, criminal, and vicarious liability.

CIVIL LAW AND CIVIL LIABILITY

No law exists that clearly stipulates when someone becomes civilly liable. Therefore, each case will have to be judged on its own merits. South African civil law provides the basic grounds on which a person can recover damages or losses suffered through the wrongful actions of another, and this is dealt with in terms of the law of delict. If someone intentionally or even just mistakenly injures someone else or damages their property that person could end up being responsible for paying for the loss of the other person. This is known as civil liability.

Regardless of the activity someone is involved in, the law requires an individual to act or behave towards other individuals in a certain, definable way, not causing them any injury or harm. The manner in which the individual must act or behave is called a standard of care. The standard of care is how the reasonable person with similar training and experience would act under similar circumstances.

The duties of care principle rests upon foreseeability. Foreseeability is part of the reasonable person test, which is applied by South African courts to establish negligence.

The reasonable person test implies the following:
1. Would a reasonable person, placed in the position of the defendant, have foreseen the possibility of causing harm or loss?
2. Would a reasonable person have taken steps to guard against the harm or loss from occurring?
3. Did the defendant take these steps?

An additional test by the South African courts is the 'reasonable expert' test and is probably more applicable to the construction industry where any number of professional experts typically form part of the construction project team. This test, although similar to the reasonable person test, should be seen as a more subjective test due to the knowledge and skills that may be expected from a person in a professional capacity such as an architect, engineer or quantity surveyor.

If the negligence of a defendant causes loss or injury to a third party or plaintiff, this person could sue for compensation in a civil court. In these cases, the State (as the State) is generally not involved. A civil action is a lawsuit filed by a private person, not the government, against another private person. Usually, these lawsuits seek monetary damages for injury or loss that the party suing or the plaintiff alleges the party being sued or the defendant caused. A defendant who loses in a civil action does not face the risk of prison or fines. A classic civil lawsuit would be a lawsuit by a homeowner against his neighbour, seeking damages in the form of money for damage to his house due to the falling tree of the neighbour.

In a civil case, the onus normally lies with the plaintiff to prove on a balance of probabilities that the defendant was negligent and caused loss or injury for which the plaintiff must be compensated. The case is heard before a judge or magistrate, depending on the damages suffered and amount claimed. Compensation is awarded by the court to the plaintiff if he has proved his case on the preponderance of probabilities.

CRIMINAL LIABILITY

In the case of criminal liability, one must differentiate between common law crimes and statutory law crimes, as shown in Table 1.1.

Table 1.1: Common law crimes compared to statutory law crimes

Common law crimes	**Statutory law crimes**
Assault	Discrimination **(Constitution)**
Culpable Homicide	Speeding **(Traffic Act)**
Murder	Tax Evasion **(Income Tax Act)**
Rape	Unhealthy Work Conditions **(OHS Act)**
Robbery	Unsafe Acts **(OHS Act)**
Theft	Unfair Dismissal **(Basic Conditions of Employment Act)**
Criminal Procedure Act	Unhealthy Work Conditions **(MHS Act)**

In criminal proceedings, prosecution will always be by the State. In cases of non-compliance with legislation, or a crime committed by an individual or business, the State prosecutes the offender, called 'the accused'. The onus lies with the State to prove beyond a reasonable doubt that the accused is guilty. The case is heard before a judge or magistrate, depending on the severity of the crime. The accused is either acquitted or convicted and sentenced.

Again, the sentence depends on the crime, and can range from a reprimand to a lifetime jail sentence and/or fine. Sometimes a criminal trial is avoided by allowing the accused to pay an admission of guilt fine as outlined in the summons. An admission of guilt remains a conviction and can have serious repercussions in a later civil action. Legal advice should therefore always be sought.

VICARIOUS LIABILITY

At times, the law imposes responsibility for civil wrongs on people or entities other than those actually engaging in the conduct that led to injury or damage. Vicarious liability relates either to criminal or civil liability and refers mainly to one party being responsible for the faults or contraventions of another.

For example, a parent could be held accountable for the loss caused through the negligence of his child, or an employer can be held accountable for a contravention by his workers, such as being held vicariously liable for theft of or damage to property at a construction site without the employer being physically present at the time, if the employer knew about the theft or damage and failed to address the issue effectively.

The concept of vicarious liability evolved from the following ancient doctrines:
- *King Hammurabi's Code of Laws – 'The Careless Supervisor':* The supervisor failed to check that the workers under his control were performing their job functions without carelessness. As a result, one worker lost his hand, and because the supervisor was guilty, one of the hands of the supervisor was also amputated.
- *Lex Aquila – 'Killing a Slave':* A slave owner was entitled to do anything with his slave, except kill him. In other words, the owner had to ensure that his slave was able to work without danger or fear of death. If the owner failed to provide the necessary protection, he was guilty of a crime.
- *Actiones Noxales – 'Sins of the Slave to be visited upon the Master':* The master was required to use his control over the slave to ensure that the slave behaved like a reasonable person. If the slave committed a crime, the master had failed to exercise the required control. This failure of control was as serious as the crime itself and the master would be responsible for the crime as if he had committed it himself.

This concept has not changed much since those early days and the supervisor today would generally be held liable for the wrongdoing of any of the workers under his supervision. Therefore, if a worker violates any provision of a statute or commits a criminal common law offence, the supervisor could be held criminally liable. And if a third party suffered loss or damage due to the negligence of the worker, he could lodge a civil claim against the supervisor. In both cases, the supervisor is liable as if he himself had committed the wrongdoing. Vicarious liability will however, in the majority of cases, only be invoked if a master and servant relationship exists. This aspect of the law is captured in the OHS Act in terms of which the CEO of a construction company would be held responsible for the occupational health and safety of the workers and be liable as if he himself had committed the wrongdoing.

1.11.2 STRICT LIABILITY

Strict liability is where a person could be held liable without any fault or negligence on his or her side. This form of liability is very seldom imposed on a person. A typical example of this kind of liability is where civil claims have been laid against Ford Motor

Company for accidents in which the drivers or owners of the Ford Kuga model have died or were seriously injured. Although the accidents were due to faulty electrical system supplied and installed by another party and not because of any fault from the side of Ford Motor Company *per se*, they are liable in terms of strict liability. The argument was that the product in its entirety was purchased by the owner and not the individual parts that make up the vehicle.

1.11.3 REASONABLY PRACTICABLE

The concept or legal doctrine of 'reasonably practicable' is a common one and is almost always captured in legal contractual documents in construction. In fact, it is the same as the reasonable person principle found in common law, but with a few specific guidelines the employer has to consider, which include:
- The scope and the severity of the hazard. A hazard could be regarded as exposure to anything that can injure a construction worker or affect the health of such a worker. The scope refers to the number of workers likely to be exposed to the hazard and the severity refers to the degree to which they will be affected.
- The means and methods to eliminate, mitigate or to reduce the effect of exposure to the hazard. These are exactly where the employer has to think like a reasonable person and implement the most appropriate measures, which might differ from situation to situation.
- The suitability of corrective measures. Again, the suitability might differ from employer to employer. The suitability will be measured against what the reasonable employer under the specific circumstances would have done. That might include aspects like culture, level of education, attitude, honesty, respect, willingness to learn, and cooperativeness of workers.
- The cost implication of the corrective measures. Some people in construction still think they can use the 'cost' versus 'benefit' argument which does not hold any water and never has, because who is able to determine the value of a life in monetary terms? That does not mean that employers do not have the right to say that there are no funds or budget available to implement certain measures. They have that right, but if that is the situation, then they either must implement alternative measures which they can afford, or stop the worker from being exposed to the hazards in question.

One of the more confusing aspects of the OHS Act is the term 'competent person'. The reason for this confusion is that it is used with different intentions and in different contexts. A wise delegate attending an H&S training workshop remarked that being competent was knowing when to ask for or seek advice.

At the outset, it is important to point out that there are two competent persons referred to in the Act, namely:
1. the dictionary meaning of a competent person; and
2. a competent person as defined in various regulations.

Although various dictionaries define the term 'competent' in a number of different ways, one can reasonably assume that the thread running through these definitions are words such as 'able, skilful, properly qualified, proper and suitable'. All persons involved in one or another construction activity, or who render a construction or professional service, have to be competent to do so. There are expectations of the abilities of a so-called 'competent person'. For example, one expects that an engineer would know what size beams and slabs to include in the design based on their expertise and professional training. The legislature therefore expects that employers will ensure that only such persons as can demonstrate the requisite expertise, qualifications and experience to carry out certain activities, actually do so. This is only relevant in those instances where the term 'competent person' is not defined in regulation 1 of the regulations in question.

According to the Department of Employment and Labour, a 'competent person' is defined as someone who has the required knowledge, training, and experience in a specific field and where applicable the relevant qualifications specific to that specific field which in this case is construction. Furthermore, this person should be familiar with the OHS Act and the applicable regulations under the Act. A competent person must demonstrate the appropriate level of competence that is appropriate for the complexity of the work or task.

For someone to be regarded as competent in construction H&S, they will at least:
- be qualified through knowledge, training and experience, and where applicable a formal qualification to do the assigned work or tasks
- have knowledge about the hazards and risks associated with the work or tasks to be performed
- know how to recognise, evaluate and control these hazards and risks
- have knowledge of the laws and regulations that apply to the work or tasks.

Knowledge can be defined as knowing both what to do as well as how to do it. Skill refers to having the ability to perform the activity correctly in terms of technical know-how, expertise, practice, measurement and feedback. Formal qualifications are earned through a formal education programme or training course or both education and practical experience.

The abilities to satisfy competency can be learned through a combination of the person's knowledge, skills and experience.

REVIEW QUESTIONS

1. Had there not been any construction H&S legislation in South Africa, how would construction workers still be protected?
2. What is the difference between common and statutory law?
3. Who is responsible for the burden of proof in a civil case and in a criminal case?
4. Explain the legal precept of *stare decisis*.
5. What are the differences between the various types of liability?

6. What is a competent person?
7. How can competence be acquired?

REFERENCES

Bamford, M. 2013. *Work and Health*. New York, NY: Springer

Bosman, AB. nd. *Manufacturers Liability and the Contingency Fees Act*. [online] Saimh. co.za. Available at: http://www.saimh.co.za/beltcon/beltcon10/paper1015.html

Department of Agriculture, Forestry and Fisheries. 1998. National Forests Act 84 of 1998

Department of Environmental Affairs. 1965. Atmospheric Pollution Prevention Act 45 of 1965

Department of Environmental Affairs. 1989. Environmental Conservation Act 73 of 1989

Department of Environmental Affairs. 1998. National Environmental Management Act 107 of 1998

Department of Labour. 1973. Occupational Diseases in Mines and Works Act 78 of 1973

Department of Labour. 1993. Compensation for Occupational Injuries and Diseases Act 103 of 1993

Department of Labour. 1993. Occupational Health and Safety Act 85 of 1993

Department of Labour. 1995. Labour Relations Act 66 of 1995

Department of Labour. 1997. Basic Conditions of Employment Act 75 of 1997

Department of Labour. 1998. Employment Equity Act 55 of 1998

Department of Mineral Resources. 1996. Mine Health and Safety Act 29 of 1996

Department of Safety and Security. 2003. Explosives Act 15 of 2003

Department of Water Affairs. 1998. National Water Act 36 of 1998

Department of Water and Sanitation. 1997. Water Services Act 108 of 1997

Gibson, C & Flood, P. 2015. *Everyone's Guide to Labour Law*. Cape Town: Penguin Random House South Africa

Haupt, TC. 2021. *Management of Safety, Health and Environment in South Africa: A Handbook*. Newcastle upon Tyne: Cambridge Scholar Publishing

Olivier, B. nd. *Corporate Governance*. [online] Financialmarketsjournal.co.za. Available at: http://financialmarketsjournal.co.za/oldsite/8thedition/printedarticles/ corporategovernance.htm

Penelope.uchicago.edu. nd. *LacusCurtius • Roman Law — Noxalis Actio (Smith's Dictionary, 1875)*. [online] Available at: http://penelope.uchicago.edu/Thayer/E/Roman/Texts/ secondary/SMIGRA*/Noxalis_Actio.html

Republic of South Africa. 1996. Constitution of the Republic of South Africa, 1996

Roman Law II - Lex Aquilia Notes. nd. *Roman Law Lex Aquilia*. [online] Available at: https://www.studocu.com/en/document/anglia-ruskin-university/law/summaries/ roman-law-ii-lex-aquilia-notes/1788700/view

Shaw, G. 2010. *RM Model for the Tourism Industry | Risk | Tourism*. [online] Scribd. Available at: https://www.scribd.com/document/188917955/RM-Model-for-the- Tourism-Industry

The laws of Hammurabi 1792 BCE to c. 1750 BCE. 1903. *The Laws of Hammurabi, King of Babylonia - Wikisource, the Free Online Library*. [online] En.wikisource.org. Available at: https://en.wikisource.org/wiki/The_Laws_of_Hammurabi,_King_of_Babylonia

Union of South Africa (U of SA). 1956. Water Act, No. 54, in U of SA, Statutes of the Union of South Africa 1956. Part II Nos. 47–73. Cape Times Ltd, under supervision of the Government Printer, Parow, pp 1046–1305

Van der Luit-Drummond, J. 2013. *The Common Law: A Student's Tale* [online] JvdLD. Available at: https://jvdld.com/2013/02/07/the-common-law-a-students-tale/

Work Health and Safety Codes of Practice. 2011. *Work Health and Safety Codes of Practice 2011*. [online] Legislation.gov.au. Available at: https://www.legislation.gov.au/Details/F2011L02804

CHAPTER 2
STATUTORY REQUIREMENTS: THE ACT

2.1 INTRODUCTION

In South African construction, all construction firms must satisfy as a minimum the requirements of two primary pieces of legislation and regulation, namely the Occupational Health and Safety Act 85 of 1993, as amended, and the Construction Regulations of 2014. These place very specific responsibilities and duties on all stakeholders participating in the construction process, from the inception of a project through to its demise at the end of its life through demolition. However, the recent emphasis worldwide has shifted from a paradigm of demolition to one of deconstruction so that what is taken down or out of a structure can be reused. Other Acts and regulations also have to be complied with.

2.2 THE ACT AND REGULATIONS

The Occupational Health and Safety Act 85 of 1993 (the Act), as amended, consists of 50 sections approved by the South African Parliament. The purpose of the Act is to provide for the health and safety of persons at work, or in connection with the use of plant and machinery. It also provides for the protection of persons other than persons at work from hazards arising out of or in connection with the activities at work. The Minister of Labour incorporates various regulations, on specific topics, into the Act from time to time.

Whereas the Constitution protects the rights of all South Africans in terms of health and safety, the Occupational Health and Safety Act 85 of 1993, as amended, and other relevant legislation prescribe the different responsibilities and liabilities of both the employer and the worker.

The Act replaced the old Machinery and Occupational Safety Act in 1993, which in turn replaced the older Factories Act of 1941 in 1983. As technology develops it is necessary for legislators to regularly revise and adapt legislation to keep abreast of technological developments and related issues such as presented by the Fourth Industrial Revolution (4IR). As new serious health threats arise that threaten the entire country or a particular region on a large scale, for example, the recent COVID-19 global pandemic, existing legislation and regulations may need to be reviewed, revised and adapted to cope with those threats.

2.3 THE PURPOSE OF THE ACT

The purpose of the Act is very clear, namely:
- To provide for the health and safety of persons at work and for the health and safety of persons in connection with the use of plant and machinery
- The protection of persons other than persons at work against hazards to health and safety arising from or in connection with the activities of persons at work
- To establish an advisory council for occupational health and safety; and to provide for matters connected therewith.

The Act requires the employer to bring about and maintain, as far as reasonably practicable, a workplace that is safe and without risk to the health of those who work in it. This means the employer must take into account the following:
- That the workplace is free of hazardous substances such as benzene and chlorine, and microorganisms, articles, equipment and processes that may cause injury, damage, illness or disease.
- The employer needs, for example, as a result of COVID-19, to ensure proper and regular deep sanitisation of all workplaces and surfaces, screening of all workers, suppliers, sub-contractors and visitors at entry and exit points of the workplace, enforcing social distancing of at least 1,5 m between workers and between their workplaces, insisting on compulsory wearing of masks, providing proper functional water supplies so that workers may regularly wash their hands with soap provided by the employer or hand sanitiser with alcohol content greater than 60%, restrict the number of workers on site at any time, rotate workers where 100% of the workforce cannot be on the site at the same time, keep proper registers and allow workers who can work from home to do so. These requirements form part of what has become known as the 'new normal'.
- The employer must inform workers of the dangers and hazards involved on the site, how to work healthily and safely, and provide personal protective equipment as a measure of last resort for a healthy and safe workplace.

However, it is not the sole responsibility of the employer to adhere to the Act. The workers also have responsibilities that they must carry out regarding certain regulations within this legislation. In short, this means that the worker and the employer share responsibility in a partnership regarding H&S in the construction workplace.

2.4 OBJECTIVES OF THE ACT

The purpose of the Act as stated earlier is to provide for the health and safety of persons at work or in connection with the use of plant and machinery. It further provides for the protection of persons other than persons at work from hazards arising out of or in connection with the activities at work. The Act supports the constitutional right of every worker to a safe and healthy workplace. The duty for creating and maintaining a healthy and safe workplace falls on every person in the workplace, to the degree that they have the authority and ability to do so. Whether they are the Chief Executive Officer (CEO) or the newest worker hired, everyone in the partnership between the employer and workers has a personal and shared responsibility for working together cooperatively to prevent occupational injuries, disease, illnesses and fatalities.

2.5 RESPONSIBILITIES DEFINED

The Act clearly defines the responsibilities of employers and workers.

2.5.1 CONSTRUCTION EMPLOYER RESPONSIBILITIES

It is commonly accepted that management of construction enterprises has the following three main areas of responsibility:

1. Financial
   - To maximise profits and minimise losses to sustain their construction-related business.
   - To prevent accidents and incidents on their construction projects, which are a major loss factor in construction firms.
2. Legal
   - To ensure that the construction firm complies with all applicable legislation, for example, the Occupational Health and Safety Act 85 of 1993 and the Compensation for Occupational Injuries and Diseases Act 130 of 1993.
3. Moral
   - To ensure that all construction workers in the firm are safe and healthy on construction sites and in construction workshops and yards so they can return home in as good a condition as when they left to come to work.

2.5.2 CONSTRUCTION HEALTH AND SAFETY MANAGEMENT FUNCTIONS

The level of management commitment and involvement in the construction H&S effort has been well documented to enhance the overall H&S performance of construction firms. It is possible to be committed but not involved. It is when both commitment and involvement are present that management becomes influential. Therefore:

- Supervisors and foremen represent the CEO of a construction company as provided for in the Act and therefore their role in managing health and safety on construction projects is not just something they do. Construction H&S forms one of their core or main duties.
- Under the Act, every employer, supervisor and foreman must, as far as possible, ensure the health and safety of all their construction workers.
- They must ensure that health and safety is managed and that all workplace processes and activities on construction projects are included in the health and safety management programme of the construction firm.

- Their leadership, ongoing commitment to and involvement in construction health and safety can play an important role in positively motivating and encouraging workers to improve health and safety on their construction projects and sites. Rules are therefore necessary and compliance to these rules is very important. Line managers such as supervisors and foremen who do not enforce these rules by disciplining and censuring workers who violate them are neglecting their duties.

Some see management as a continuous cycle of planning, organising, leading and controlling the efforts of others in order to achieve specific goals. Management is a process aimed at the realisation of objectives or goals through the utilisation of the efforts of others. The goal of construction firms is to construct facilities, manufacture or supply goods or services in such a way that a reasonable profit can be earned to cover the costs of being in business and sustain the longevity of the firm itself. Accidents/incidents are costly and can therefore have an adverse effect on the profitability of any construction enterprise. To make a profit, management must exercise control in such a way that accidents and incidents on construction sites are kept to a minimum to prevent the effects of their adverse consequences on the finances of the organisation.

The employer must make sure that the construction site, workshop or yard is safe and healthy and must not allow any worker to do work that is potentially dangerous.

CHIEF EXECUTIVE OFFICER (CEO)

The chief executive officer (CEO) of a construction firm represents the employer and is as such personally responsible for the health and safety of each construction worker as well as to ensure that all the employer duties stipulated by the Act are met. Therefore, CEOs can be held personally liable should they not execute their responsibilities towards all their workers.

In terms of section 16 of the Act:
- The CEO must ensure the proper execution of the duties of the employer in terms of the Act.
- The CEO may assign duties in writing to any person under his control, who shall act subject to the control and directions of the CEO.
- The head of department of any department of State such as the Departments of Public Works and Infrastructure; Health; and Education shall be deemed to be the chief executive officer of that department.

In practice, the CEO will assign in terms of section 16(2) specific functions to various line managers such as site agents, project managers, supervisors and foremen and appoint specific staff in the organisation to assist with this duty.

However, the following still applies:
- Delegation never means that the final responsibility is delegated. The CEO stays personally responsible and liable for all the regulations as stipulated in the Act, even if the CEO has delegated certain functions to someone else.
- Staff to whom specific construction health and safety duties have been assigned will perform these duties under the control of the CEO. Moving beyond duties and responsibilities is part of that unwritten function.

Health and safety on construction projects should go beyond compliance with legal provisions and actions aimed primarily at avoiding prosecution. It is also a moral issue that reflects on the ethical values of the construction firm and has economic and cost implications for both the firm itself as well as the construction worker. It is a construction project issue that should be dealt with through constructive participation and joint problem solving between the relevant role players in the organisation. Construction health and safety should not be a priority of the firm but an organisational value and part of a positive organisational health and safety culture and climate.

ROLE OF SUPERVISORS
Generally, supervisors and foremen aim at ensuring that production targets on construction projects and sites, as set out in the project schedule or programme, are met as their primary role in a construction firm. However, the Act does not concern itself with the production output of the employer, but rather with the health and safety of workers and others that may be affected by the activities of the employer. Ensuring adherence to these standards cannot only be placed on the shoulders of management. Further assistance is required, and supervisors and foremen perform this role in the overall construction health and safety effort of the organisation.

In terms of section 8(2)*(e)* and *(i)* of the Act, the role of supervisors and foremen is to:
1. provide such information, instructions, training and supervision as may be necessary to ensure, as far as is reasonably practicable, the health and safety at work of workers
2. ensure that work is performed and that plant and machinery are used under the general supervision of a person trained to understand the hazards associated with them and who has the authority to ensure that precautionary measures taken by the employer are implemented.

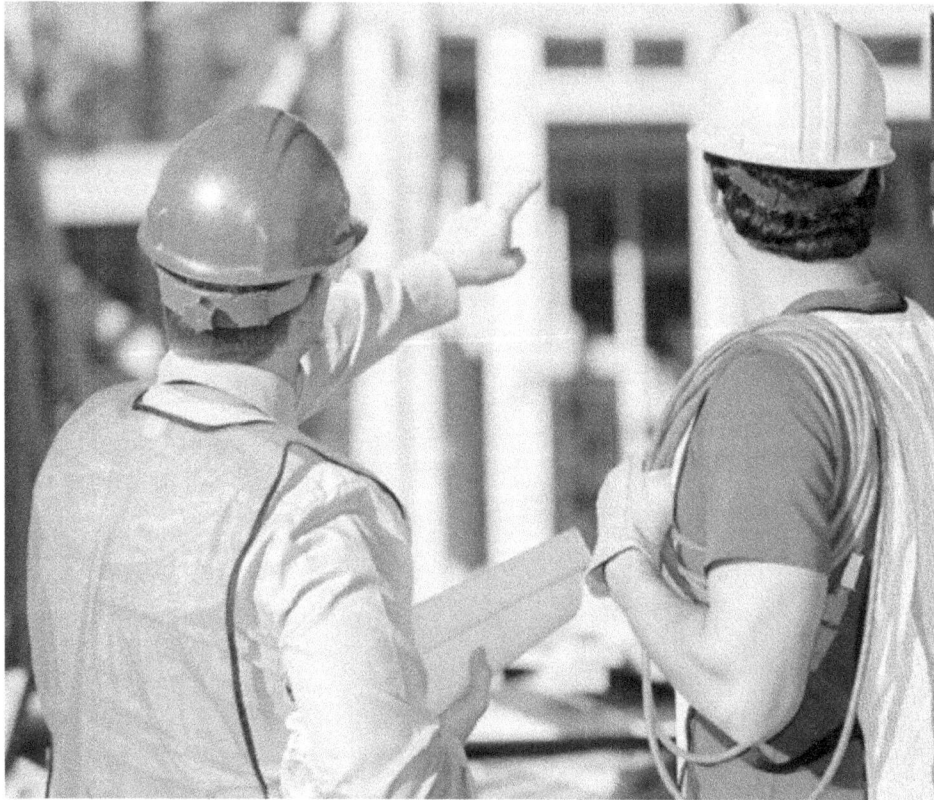

Figure 2.1: Importance of the role and influence of the supervisor on a project

2.5.3 GENERAL DUTIES OF CONSTRUCTION EMPLOYERS TO THEIR WORKERS

In terms of section 8 of the Act, all employers shall provide and maintain, as far as is reasonably practicable, a working environment that is safe and without risk to the health of their workers.

The matters to which these duties refer include the following:

- The provision and maintenance of systems of work, plant and machinery that, as far as is reasonably practicable, are safe and without risks to health. The employer must provide and maintain all the equipment that is necessary to do the work, and all the systems according to which work must be done, in a condition that will not affect the health and safety of workers.
- Taking steps to eliminate or minimise any hazard or potential hazard to the safety or health of workers, before resorting to the provision and use of personal protective equipment. Before personal protective equipment may be used, the employer must first conduct a risk assessment and then mitigate exposure to any residual hazards that may affect the health and safety of his workers. Only when this is not practical,

should personal protective equipment be provided and used. Therefore, the use of personal protective equipment should always be a measure of last resort to, inter alia, eliminate giving the worker a false sense of security that they are safe and without any threat to their health and well-being.

- Ensuring the safety and absence of risks to health in connection with the production, processing, use, handling, storage or transport of articles or substances, including all workplaces and surfaces. In other words, anything that workers may come into contact with at work.
- Establishing what hazards to the health or safety of workers and the public exist, what precautionary measures should be taken with respect to such hazards, and providing the necessary means to apply such precautionary measures.
- Providing information, instructions, training and supervision necessary to ensure the health and safety at work of all workers.
- Not permitting any worker to do any work unless the prescribed precautionary measures have been taken.
- Taking all necessary measures to ensure that the requirements of this Act are complied with by every worker in his employment or on premises under his control where plant or machinery is used.
- Enforcing such measures in the interest of health and safety.
- Ensuring that work is performed under supervision of a person who will ensure that precautionary measures taken by the employer are implemented.
- Informing workers regarding the scope of their authority. (Refer to section 37 of the Act.)

To ensure that these duties are complied with, the employer must do the following:
- Through conducting a Hazard Identification and Risk Assessment (HIRA), identify any potential residual hazards which may still be present while construction work is being carried out, a product is being produced, processed, used, stored or transported, and any construction equipment and tools are being used after mitigating interventions to reduce the level of exposure have been considered and introduced.
- Establish the precautionary measures that are necessary to protect construction workers against the identified hazards and provide the means to implement these precautionary measures.
- Provide the necessary information, instructions, training and supervision while keeping the extent of the competence of construction workers in mind; in other words, what they may do and may not do.
- Not permit anyone to carry on with any task unless the necessary precautionary measures have been taken.
- Take steps to ensure that every person under the control of the employer or construction firm complies with the requirements of the Act.
- Enforce the necessary control measures in the interest of construction health and safety.

- Ensure that the work being done, and the equipment being used are under the general supervision of a construction worker who has been trained to understand the hazards associated with the construction work. Such a worker must ensure that the precautionary measures are implemented and maintained.

ARTICLES USED IN CONSTRUCTION

Manufacturers, designers, importers, sellers and suppliers must ensure that the following are adhered to regarding articles used in construction:

- Their articles are safe and without risk to health and comply with all prescribed requirements.
- When a structure or an article is installed on any premises, it must be done in such a way that neither an unsafe situation nor a health risk is created.

SUBSTANCES USED IN CONSTRUCTION

Similarly, manufacturers, designers, importers, sellers and suppliers must ensure that the following is adhered to regarding substances use in construction:

- Such substances are safe and without risk to the health and safety of construction workers when properly used.
- Information is available on the following:
  - Use of the substance at the construction workplace
  - Health and safety risks associated with the substance
  - Conditions that are necessary to ensure that the substance will be safe and without risk to the health of construction workers when properly used
  - Procedures in case of an accident or incident on the construction site are in place.

Usually, this information is contained in a Material Safety Data Sheet or MSDS, which should be kept on the construction site and be available for consultation and training in the proper use of the articles and substances. While these documents of information must be kept in the project health and safety file, copies should also be available where the relevant construction activities are being executed by the workers. If a party, in this case a construction firm, to whom an article or substance has been sold or supplied, undertakes in writing to take specified steps to ensure that the article or substance will meet all the prescribed requirements, and will be safe and without risk to health, the duties of the importer, designer, seller, supplier or manufacturer will subsequently shift to the party who undertakes to take such steps.

DUTY OF EMPLOYERS TO INFORM THEIR EMPLOYEES: SECTION 13

The employer must ensure that every construction worker is informed and clearly understands the health and safety hazards on the construction site or any construction work that has to be done, anything being produced, processed, used, stored, handled or transported, and any equipment or machinery being used. The employer must then provide full and detailed information about precautionary measures against any likely exposure to these hazards.

The employer must inform health and safety representatives when an inspector notifies the employer of inspections and investigations to be conducted at the construction site or plant. The employer must also inform health and safety representatives of any application for exemption made, or of any exemption granted to him or her in terms of the Act. Exemption means being exempted from having to comply with certain provisions of the Act, regulations, notices or instructions issued under the Act.

The employer must, as soon as possible, inform the health and safety representatives of the occurrence of an incident or accident on the construction site. An incident or accident is an event that occurs in the workplace where a person is killed, injured or becomes ill. It also includes the spillage of a hazardous substance, for example, where a tank leaks diesel, which is commonly used on construction sites, due to a faulty valve, or where construction equipment such as a plate compactor runs out of control, without necessarily killing or injuring anyone.

PROVISION BY EMPLOYERS OF INFORMATION AND TRAINING ABOUT HAZARDS
Employers must do the following:
• Inform workers of the dangers in the construction workplace such as a construction site, workshop or yard.
• Ensure that there are warnings and notices on all dangerous construction machinery.
• Train construction workers on how to use dangerous machinery or substances safely.
• Ensure that all construction equipment is properly maintained.
• Provide appropriate protective clothing and equipment where necessary.
• Reduce any dangers to a minimum before issuing protective clothing.
• Make sure that a supervisor oversees operations and construction activities and enforces safety requirements.
• Set out precautionary measures to prevent dangers.

2.5.4 LIABILITIES OF EMPLOYERS

In terms of section 38(2) of the Act:

> Any employer who does or omits to do an act, thereby causing any person to be injured at a workplace, or, in the case of a person employed by him, to be injured at any place in the course of his employment, or any user who does or omits to do an act in connection with the use of plant or machinery, thereby causing any person to be injured, shall be guilty of an offence if that employer or user, as the case may be, would in respect of that act or omission have been guilty of the offence of culpable homicide had that act or omission caused the death of the said person, irrespective of whether or not the injury could have led to the death of such person, and on conviction be liable to a fine not exceeding R100 000 or to imprisonment for a period not exceeding two years or to both such a fine and such imprisonment.

2.5.5 CONSTRUCTION WORKER RESPONSIBILITIES

In terms of section 14 of the Act, it is the duty of every construction worker on a construction site, plant or workshop to ensure that he or H&S complies with the following:

1. Take reasonable care for the health and safety of himself and of other persons who may be affected by his acts or omissions. This includes playing at work. Many people have been injured and even killed due to horseplay in the workplace, and that is considered a serious contravention.
2. Cooperate with his employer with regards to any duty or requirement imposed on his employer by this Act, to enable that duty or requirement to be performed or complied with.
3. Carry out any lawful instruction which the employer or authorised person prescribes about health and safety. Comply with the rules and procedures that the employer gives him or her.
4. Wear the prescribed safety clothing or the prescribed safety equipment where required.
5. Report any situation which is unsafe or unhealthy to his employer or to the health and safety representative for his workplace or section thereof, who shall report it to the employer.
6. Report any incident that may affect his health or that has caused an injury to himself to his employer or to his health and safety representative, not later than the end of the particular shift during which the incident occurred, unless the circumstances were such that the reporting of the incident was not possible, in which case he shall report the incident as soon as practicable thereafter.
7. Give information to an inspector from the Department of Labour if he or she should require it.

DUTY NOT TO INTERFERE WITH, DAMAGE OR MISUSE THINGS

In terms of section 15 of the Act: 'No person shall intentionally or recklessly interfere with, damage or misuse anything which is provided in the interest of health or safety.' For example, no part of a scaffold that is in use should be removed that will cause it to become structurally deficient and affect its structural integrity.

2.5.6 THE RIGHTS OF CONSTRUCTION WORKERS

The Act has extended the rights of construction workers to include the following:

1. The right to information
2. The right to participate in inspections
3. The right to comment on legislation and make representations
4. The right not to be victimised
5. The right to appeal.

THE RIGHT TO INFORMATION

Construction workers must have access to the following:
- The Act and associated regulations under the Act
- Health and safety rules and procedures of the organisation and construction site, plant or workshop
- Health and safety standards of the organisation and construction site, plant or workshop that must be kept by the employer.

The worker[1] may request the employer to inform him or her regarding the following:
- Health and safety hazards on the construction site or in the plant or workshop
- The precautionary measures which must be taken
- The procedure that must be followed if a construction worker is exposed to substances hazardous to their health.

Construction workers may request that their private medical practitioner or doctor investigate their medical and exposure records. If the workers are also health and safety representatives, they may investigate and comment in writing on exposure assessments and monitoring reports.

THE RIGHT TO COMMENT ON LEGISLATION AND MAKE REPRESENTATIONS

Construction workers may comment or make representations on any regulation or safety standard published under the Act.

THE RIGHT NOT TO BE VICTIMISED

An employer may not dismiss a construction worker[2] from his service, reduce the salary of a construction worker or reduce the service conditions of a construction worker because:
1. The worker supplied information, which is required of him or her in terms of the Act, to someone who is charged with the administration of the Act.
2. The worker complied with a lawful notice such as a prohibition or a contravention notice.
3. The worker did something that should have been done in terms of the Act.
4. The worker did not do something that is prohibited in terms of the Act.
5. The workers have given evidence before the Labour Court or a court of law on matters regarding health and safety.

THE RIGHT TO APPEAL

The worker may appeal against the decision of an inspector by making a written submission to the chief inspector.

---

1   The singular form applies to both male and female workers.
2   A construction worker as used in this text refers to any person employed by the construction firm.

DUTY NOT TO INTERFERE WITH OR MISUSE OBJECTS

No one may interfere with or misuse any object that has been provided in the interest of health and safety. A person may, for example, not remove a safety guard from a machine and use the machine or allow anybody else to use it without such a guard.

2.6   DEPARTMENT OF LABOUR IN THE ACT

The Chief Directorate of Occupational Health and Safety of the Department of Labour administers the Act.

In order to ensure the health and safety of construction workers across South Africa, provincial offices have been established in all nine provinces. To this end, Occupational Health and Safety inspectors from these provincial offices carry out inspections and investigations on the construction site or in the plant or workshop.

Inspections are usually planned on the basis of accident statistics, the presence of hazardous substances, such as the use of hazardous chemical substances as additives to cement or concrete, or the use of dangerous plant and equipment on construction sites. Unplanned inspections, on the other hand, usually arise from requests or complaints by workers, employers or members of the public. These complaints or requests are treated confidentially.

2.6.1 INSPECTORS

In terms of section 28 of the Act:
1.   The Minister may designate any person as an inspector to perform the functions assigned to an inspector by this Act.
2.   Each designated inspector must have a certificate signed by or on behalf of the Minister, stating that he has been designated as an inspector.
3.   Inspectors may be required to produce such a certificate on demand during performance of their functions under this Act.

In terms of section 29 of the Act, an inspector may, for the purposes of this Act:
1.   Without previous notice, at all reasonable times enter any construction site, plant, store or workshop.
2.   Question any person in the workplace on any matter to which this Act relates.
3.   Require from any person in possession or control of a book, record or other document on or in the workplace, to produce to him such book, record or other document.
4.   Examine any such book, record or other document or make a copy thereof or an extract therefrom.

5. Require from such a person an explanation of any entry in such book, record or other document.
6. Inspect any article, substance, plant, or machinery which is used in the workplace, or any work performed, or any condition prevalent in the workplace. He may also remove for examination or analysis any article, substance, plant or machinery or a part or sample thereof.
7. Seize any such book, record, or other document or any such article, substance, plant or machinery or a part or sample thereof which in his opinion may serve as evidence at the trial of any person charged with an offence under this Act or the common law.
8. The employer may make copies of documents before such seizure.
9. Direct any employer, employee or user, including any former employer, employee or user, to appear before him on any matter to which this Act relates.
10. Perform any other function as may be prescribed.

An interpreter, a member of the South African Police or any other assistant may, when required by an inspector, accompany him when he performs his functions under this Act. For the purposes of this Act an inspector's assistant shall, while he acts under the instructions of an inspector, be deemed to be an inspector.

When an inspector enters any construction site, workshop, yard or store the employer and construction workers performing any work must at all times provide the inspector with facilities to enable him and his assistant (if any) to perform effectively and safely his or their functions under this Act.

When an inspector removes or seizes any article, substance, plant, machinery, book, record or other document he must issue a receipt to the owner or person who is in control of it.

SPECIAL POWERS OF INSPECTORS
In terms of section 30 of the Act: If an inspector finds dangerous or adverse conditions on any construction site, in any plant, store or workshop, he or she may set requirements to the employer in the following ways and serve the following documents on the employer:
• Prohibition notice
• Contravention notice
• Improvement notice
• Other powers.

Prohibition notice
In the case of threatening danger an inspector may prohibit a particular action, process, or the use of a machine or equipment, by means of a prohibition notice. No person may disregard the contents of such a notice and compliance must take place with immediate effect.

Contravention notice

If a provision of a regulation is contravened, the inspector may serve a contravention notice on the construction workers or the employer. A contravention of the Act can result in immediate prosecution, but in the case of a contravention of a regulation, the employer may be given the opportunity to correct the contravention within a limit specified in the notice, which is usually 60 days.

Improvement notice

Where the construction health and safety measures that the employer has instituted do not satisfactorily protect the health and safety of the workers, the inspector may require the employer to bring about more effective measures. An improvement notice, which prescribes the corrective measures, is then served on the employer.

Non-compliance notice

A non-compliance notice directs an employer to appear at an enquiry to investigate the alleged non-compliance with the requirements of the Act.

Other powers

To enable the inspector to carry out his or her duties, he or she may enter any construction site, plant, store or workshop where machinery or hazardous substances are being used and question or serve a summons on persons to appear before him or her. The inspector may request that any documents be submitted to him or her, investigate and make copies of the documents, and demand an explanation about any entries in such documents. The inspector may also inspect any condition or article and take samples of it and seize any article that may serve as evidence.

Should an employer not be happy with the actions of the inspector, he may appeal to the Chief Inspector in writing within 60 days.

> *Note: The powers of inspectors are not absolute. Any person who disagrees with any decision taken by the inspector may appeal against that decision in writing to the Chief Inspector Occupational Health and Safety at the Department of Labour.*

COOPERATION WITH THE INSPECTOR

Compliance with directions, subpoenas, requests or commands

Employers and construction workers must comply with the directions, subpoenas, requests or orders of inspectors. In addition, no one may prevent any other worker or anyone else from complying.

Answering questions

The questions posed by inspectors should be answered, but no one is obliged to answer a question by which he or she might incriminate him or herself. To incriminate oneself means that one is suggesting that one is responsible for a contravention.

Investigations

When the inspector so requires, he or she must be provided with the necessary means and be given the assistance he or she may need to hold an investigation. The inspector may also request that individuals attend investigations that may assist the inspector with the investigation. No one may insult the inspector or deliberately interrupt the investigation.

Prosecutions

When the construction worker does something which in terms of the Act is regarded as an offence, the employer is responsible for that offence as explained previously in terms of vicarious liability, and he or she could be found guilty and sentenced for it, unless the employer can prove that:

- he or she did not give his or her consent
- he or she took all reasonable steps to prevent it from happening
- the construction worker did not act within the scope of his or her competence, in other words, that the worker did something which he or she knew he or she should not have done.

2.6.2 CONSTRUCTION HEALTH AND SAFETY REPRESENTATIVES

In terms of section 17 of the Act:

1. Employers, who employ more than 20 construction workers at any construction site, plant, store, or workshop,[3] must appoint health and safety representatives in writing for such workplace, or for different sections thereof.

2. An employer and the construction worker representatives, or workers where there are no such representatives, must consult in good faith regarding the election, period of office and subsequent designation of health and safety representatives. If such consultation fails, the matter shall be referred for arbitration to a person mutually agreed upon, whose decision shall be final.

3. Only full-time construction workers, who are acquainted with conditions and activities at that construction site, plant, store or workshop, will be eligible for designation as construction health and safety representatives for that construction site, plant, store or workshop.

4. The number of construction health and safety representatives for a construction site, plant, store or workshop or section thereof shall be at least one construction health and safety representative for every 50 workers or part thereof. Workers working at a construction site, plant, store or workshop other than that where they ordinarily report for duty, shall be deemed to be working at the construction site, plant, store or workshop where they report for duty.

---

3   Where the term 'workplace' has been used in the wording of the Act, it has been substituted with 'construction site, plant, store or workshop'.

5. Inspectors of the Department of Labour (DoL) may direct employers in writing to increase the number of construction health and safety representatives if they feel that, under certain circumstances, the prescribed number is not adequate.

6. All activities in connection with the activities and training of construction health and safety representatives shall be performed during ordinary working hours, and any time reasonably spent by any employee in this regard shall for all purposes be deemed to be time spent by him in the carrying out of his duties as a worker.

Irrespective of whether the number of workers employed at the construction firm is fewer than 20 workers, which technically does not require there to be a construction H&S representative, there should always be construction H&S representatives. There should be at least one construction H&S representative on each site that has fewer than 20 workers and more construction H&S representatives on larger projects to ensure that the health and safety issues and concerns of all stakeholders are adequately addressed. Bear in mind what the intent of having construction H&S representatives is rather than blindly trying to satisfy the requirements of the Act, which are the minimum in any case. In the construction sector, construction H&S representatives should be employed on a full-time basis in that capacity with a job description befitting their appointment, and be empowered through appropriate training to perform the functions currently listed for construction H&S officers by the South African Council for the Project and Construction Management Professions (SACPCMP).

FUNCTIONS OF CONSTRUCTION HEALTH AND SAFETY REPRESENTATIVES

In terms of section 18 of the Act: A construction health and safety representative may perform the following functions in respect of the workplace or section of the workplace for which he has been designated:

1. Review the effectiveness of construction health and safety measures.
2. Identify potential hazards and major incidents at the construction site, plant, store or workshop.
3. Examine the causes of incidents at the construction site, plant, store or workshop.
4. Investigate complaints by any employee relating to the health or safety at work of that employee.
5. Make representations to the employer or a construction health and safety committee on matters arising from these aspects.
6. Make representations to the employer on general matters affecting the health or safety of the employees at the construction site, plant, store or workshop.
7. Inspect the construction site, plant, store or workshop with a view to the health and safety of employees, at times agreed upon with the employer, who may be present during the inspection. The employer must be given reasonable notice of such inspections.
8. Participate in consultations with inspectors at the construction site, plant, store or workshop and accompany inspectors on inspections of the workplace.

9. Receive information from inspectors as contemplated in section 36 of the Act.
10. Attend meetings of the construction health and safety committee of which he is a member, in connection with any of these functions.

2.6.3 CONSTRUCTION HEALTH AND SAFETY COMMITTEES

In terms of section 19 of the Act:
1. Employers must establish one or more construction health and safety committees where two or more construction health and safety representatives have been designated, and, at every meeting of such a committee, consult with the committee on measures to ensure the health and safety of his employees at work.
2. Nominees on a construction health and safety committee shall be designated in writing by the employer for a specific period. Construction health and safety representatives must be members of the committee for the period of their designation.
3. Construction health and safety committee meetings must be held at least once every three months, at a time and place determined by the committee. Under certain circumstances a DoL inspector may in writing direct a construction health and safety committee to hold a meeting.
4. A construction health and safety committee may appoint advisory members with particular knowledge of construction health or safety matters. These advisory members will have no voting rights.
5. Inspectors of the DoL may direct employers in writing to increase the number of construction health and safety committees if they feel that, under certain circumstances, the prescribed number is not adequate.

It is important that there is a construction health and safety committee in every construction firm irrespective of the number of construction workers that are employed, or number of H&S representatives, so that construction H&S can be elevated and receive the attention that it should and so that the overall H&S performance of the firm together with the safety, health and well-being of everyone can be improved.

FUNCTIONS OF CONSTRUCTION HEALTH AND SAFETY COMMITTEES: THE ACT
In terms of section 20 of the Act:
1. A construction health and safety committee may make recommendations to the employer or an inspector regarding any matter affecting the health or safety of persons at the construction site, plant, store or workshop, and discuss and report to an inspector any incident causing injury or death at the construction site, plant, store or workshop.
2. A construction health and safety committee must keep a record of recommendations made to an employer and of any report made to an inspector.

Employers must take the prescribed steps to ensure that a construction health and safety committee complies with the provisions of the Act.

PURPOSE OF CONSTRUCTION HEALTH AND SAFETY COMMITTEES
Members of construction H&S committees meet in order to initiate, promote, maintain and review measures of ensuring the health and safety of construction workers and everyone associated with the construction firm as well as the public affected by its activities.

ESTABLISHMENT OF CONSTRUCTION HEALTH AND SAFETY COMMITTEES
At least one committee must be established when two or more representatives are designated. It is recommended that given the intent of having such a forum, there always be a construction health and safety committee irrespective of the number of construction health and safety representatives, especially considering the role that the committee plays in the overall health and safety effort of the construction firm and the current general poor H&S performance of the construction industry.

COMPOSITION OF A CONSTRUCTION HEALTH AND SAFETY COMMITTEE
The employer determines the number of construction health and safety committee members, based on the following:
- If only one committee has been established for a construction site, plant, store or workshop, all the representatives must be members of that committee.
- If two or more have been established for a construction site, plant, store or workshop, each representative must be a member of at least one of those committees.
- Therefore, every construction H&S representative must be a member of a construction H&S committee.

The employer may also nominate other persons to represent him or her on a committee but such nominees may not be more than the number of representatives designated on that committee.

If, however, an inspector is of the opinion that the number of committees in the construction site, plant, store or workshop is inadequate, he or H&S may determine the establishment of additional committees.

FREQUENCY OF MEETINGS OF CONSTRUCTION HEALTH AND SAFETY REPRESENTATIVES
The following guidelines are useful:
1. They meet whenever it is necessary, but at least once every three months.
2. The construction H&S committee determines the time and place.
3. However, if 10% or more of the construction workers put forward a request for a meeting to the inspector, the inspector may order that such a meeting be held at a time and place which he or she determines.

The members of the committee elect the chairperson and determine, for example, his or her period of office, and meeting procedures. The chairperson could also be rotated between the employer and the construction H&S representatives.

Construction H&S committees may co-opt persons as advisory members for their knowledge and expertise on construction H&S matters. For example, the occupational hygienist and nurse could be members of the committee. However, an advisory member does not have any voting powers.

FUNCTIONS OF CONSTRUCTION HEALTH AND SAFETY COMMITTEES
The Construction H&S committees only deal with health and safety matters in the construction workplace or sections of the site if it is a large one, for which such committees have been established. Generally, construction H&S committees have the following functions:

1. **Make recommendations:** A construction H&S committee must make recommendations about the health and safety of construction workers to the employer. Where these recommendations do not lead to solving the matter, the construction H&S committee may make recommendations to an inspector.
2. **Discuss incidents:** A construction H&S committee must discuss any incident that leads to the injury, illness or death of any construction worker and may report it in writing to an inspector.
3. **Record keeping**: A construction H&S committee must keep a record of every recommendation to the employer and every report to an inspector.
4. **Other functions:** Construction H&S committee members must perform any other functions required of them by regulation.

2.6.4 DEDUCTIONS

An employer may not make any deduction from the salary or wages of a construction worker with regard to anything he or she is required to do in the interest of health and safety in terms of the Act.

2.6.5 REPORT TO CHIEF INSPECTOR ABOUT OCCUPATIONAL DISEASES

If a medical practitioner examines or treats a construction worker for a disease that he or she suspects arose from the employment of the worker, the medical practitioner must report the matter to the employer of the worker and to the chief inspector, and notify the worker.

> *The Act or regulations can be purchased from the government printer in Gazette form or bound form from various publishers or downloaded from the DoL link on the government website: gov.za*

2.7 CONSTRUCTION WORK PROCEDURES

No specific section or regulation in the Act requires the employer to have written safe work procedures for every task performed by a construction worker. What the Act does require is for the employer to ensure that every construction worker is informed of the hazards and risks associated with the tasks the construction worker is required to perform.

Section 37 of the Act makes the employer liable for the acts or omissions of construction workers, unless the employer can prove the following:
(a) The construction worker acted without the permission of the employer;
(b) It was not within the scope of authority of the construction worker to do the act; and
(c) The employer took all reasonable steps to prevent the act in question.

Note that the fact that the employer issued instructions forbidding the act in question, shall in itself not be regarded as proof that all reasonable steps were taken.

Detailed safe work procedures would assist with the following:
• Documented proof of hazards identified
• Documented proof of risks identified
• Documented proof of identified precautionary measures
• Documented proof of the scope of authority of construction workers
• Immediate availability of training material
• Proof that reasonable measures were taken to ensure that construction workers followed the correct work procedures
• Assistance with the job specifications of construction workers
• Provide the standard for planned construction task or activity observations.

2.8 LEGISLATIVE PROCEDURES

Every construction organisation presents health and safety risks to both their construction workers, visitors to the construction site, plant, store or workshop and their clients. To minimise the risk of injury to their construction workers, visitors and their clients, all their workers need to acknowledge and practise certain health and safety procedures. The health and safety of construction workers is regulated by the Act and associated regulations. All construction firms need to subscribe to legislative procedures in order to ensure the minimum health and safety of their construction workers and clients. Table 2.1 lists some suggestions on the actions required from employers and workers alike.

Table 2.1: Suggestions regarding legislative procedures

	Action
1	Maintain in good condition insulated stands, trestles, mats or other such protective equipment as may be necessary to prevent accidents for use by persons working in close proximity to electrical equipment on the construction site.
2	Construction workers must take reasonable care for their own health and safety.
3	Construction workers must take steps as may be reasonably practicable to eliminate or mitigate any hazard or potential hazard to their own health and safety, as well as that of other construction workers and clients.
4	Construction workers must take precautionary measures that are prescribed.
5	Construction workers must ensure that relevant signage is visible.
6	Affix a prominent notice or sign in a conspicuous place on the construction site, indicating where the first aid box or boxes are kept as well as the name of the person in charge of such first aid box or boxes.
7	Where more than 10 workers are employed on a construction site, take steps to ensure that for every group of up to 50, a minimum of one qualified first aid representative is available at that site.

2.9 CONSTRUCTION FIRM PROCEDURES

Construction firms should have procedures in place that are implemented to suit the specific needs of the particular firm and its construction activities. There are, however, general expectations with regard to procedures concerning construction health and safety. Some examples of procedures are shown in Table 2.2.

Table 2.2: Examples of procedures concerning health and safety

	Action
1	Do not run in the workplace.
2	Wear safety footwear with non-slip soles.
3	Clean up spills immediately.
4	Put up signs to mark any wet areas.
5	Be familiar with the location of the first aid box on the construction site.
6	Make sure that construction workers know the names of the first aid and safety officers on their projects and how to contact them.
7	Have knowledge of the relevant emergency authorities and how to contact them.
8	Display emergency and caution signs on the construction site where they are easily accessible and observed.
9	Effectively display fire exits and routes on the construction site.

The primary reasons for evaluating, implementing and maintaining H&S on the construction site, in the plant, store or workshop include:

- The creation of a safe and healthy working environment which, if aimed for, makes compliance with legislation much easier;
- To create a healthy and safe system of work that ensures that hazards are identified, and risks are assessed and controlled;
- To ensure that plant and machinery are provided and maintained in a safe working order.

When the minimum requirements for construction H&S captured in legislation and regulations are not complied with the consequences can be dire and disastrous. For example:

- Construction workers are injured or killed.
- Construction workers and others can be exposed to harmful substances and be diagnosed with an occupational disease.
- Construction workers can be maimed by machinery by possibly losing a limb such as an arm, leg or hand.
- There could be fines or imprisonment for the CEO of the construction firm.
- Construction projects can be stopped or closed, which will impact severely on the financial resources of the business.
- The local community and family system are severely impacted by the potential loss of a breadwinner due to an accident on a construction site:
  - Children growing up without a father or mother due to either of them being killed or permanently disabled at the construction site, plant, store or workshop and the impact on their present and future circumstances
  - Criminal prosecution of the employer and the construction worker
  - Civil action against the employer and the construction worker
  - Increases in Compensation Commissioner or Federated Mutual Association premiums
  - Additional compensation to the injured
  - Construction firm closing down or being shut down, leading to loss of income for everyone.

2.10 CLOSING OF OPERATIONS

The following article was released a few years ago by the Department of Labour in relation to the consequences of not complying with the Occupational Health and Safety Act.

LABOUR DEPARTMENT SHUTS DOWN COMPANY, ISSUES WARNINGS

Wednesday, March 7, 2012

Pretoria – The Department of Labour closed down a company for non-compliance with the Occupational Health and Safety Act and issued several warnings to other businesses during an inspection operation in Johannesburg on Tuesday.

The department issued 24 businesses with written warnings for contravening the Basic Conditions of Employment Act, including underpaying workers. A total of 24 illegal immigrants were arrested at the firms.

The department in conjunction with the police, immigration officials, traffic cops and the South African Revenue Services (SARS) conducted a joint inspection aimed at clamping down on companies not complying with labour laws and illegal business operations.

Gauteng Department of Labour spokesperson, Mishack Magakwe, said the notices of non-compliance that were issued will be followed up on. 'We will continue to enforce labour legislation', he said.

According to the department, in some instances, employees were working without any contracts, pay slips or attendance registers and were being underpaid.

'Some of the employees were not registered with the UIF and the compensation fund while the Occupational Health and Safety Act was found to be non-existent,' Magakwe said, adding that employers were given 21 days to correct matters.

'They were issued with written undertakings to comply with, or risk being dragged to court,' he said.

The department said more joint inspections with SAPS, Home Affairs, SARS and Metro Police in the province could be expected.

(Source: https://www.sanews.gov.za/south-africa/labour-dept-shuts-down-company-issues-warnings)

2.11 NON-COMPLIANCE: EMPLOYEE AND EMPLOYER LIABILITIES

The following is an extract from the Act, as amended, about the consequences of non-compliance with its requirements:

Acts or omissions by employees or mandataries

37. (1) Whenever an employee does or omits to do any act which would be an offence in terms of this Act for the employer of such employee or a user to do or omit to do, then, unless it is proved that –

    (a) in doing or omitting to do that act the employee was acting without the connivance or permission of the employer or any such user;

    (b) it was not under any condition or in any circumstance within the scope of the authority of the employee to do or omit to do an act, whether lawful or unlawful, of the character of the act or omission charged; and

    (c) all reasonable steps were taken by the employer or any such user to prevent any act or omission of the kind in question,

    the employer or any such user himself shall be presumed to have done or omitted to do that act, and shall be liable to be convicted and sentenced in respect hereof; and the fact that he issued instructions forbidding any act or omission of the kind in question shall not, in itself, be accepted as sufficient proof that he took all reasonable steps to prevent the act or omission.

(2) The provisions of subsection (1) shall mutatis mutandis apply in the case of a mandatory of any employer or user, except if the parties have agreed in writing to the arrangements and procedures between them to ensure compliance by the mandatory with the provisions of this Act.

(3) Whenever any employee or mandatory of any employer or user does or omits to do an act which it would be an offence in terms of this Act for the employer or any such user to do or omit to do, he shall be liable to be convicted and sentenced in respect thereof as if he were the employer or user.

(4) Whenever any employee or mandatory of the State commits or omits to do an act which would be an offence in terms of this Act, had he been the employee or mandatory of an employer other than the State and had such employer committed or omitted to do that act, he shall be liable to be convicted and sentenced in respect thereof as if he were such an employer.

(5) Any employee or mandatory referred to in subsection (3) may be so convicted and sentenced in addition to the employer or user.

(6) Whenever the employee or mandatory of an employer is convicted of an offence consisting of a contravention of section 23, the court shall, when making an order under section 38(4), make such an order against the employer and not against such employee or mandatory.

Offences, penalties and special orders of court

38. (1) Any person who –

[...]

(d) in any record, application, statement or other document referred to in this Act wilfully furnishes information or makes a statement which is false in any material respect;

(e) hinders or obstructs an inspector in the performance of his functions;

(f) refuses or fails to comply to the best of his ability with any requirement or request made by an inspector in the performance of his functions;

(g) refuses or fails to answer to the best of his ability any question which an inspector in the performance of his functions has put to him;

(h) willfully furnishes to an inspector information which is false or misleading;

(i) gives himself out as an inspector;

(j) having been subpoenaed under section 32 to appear before an inspector, without sufficient cause (the onus of proof whereof shall rest upon him) fails to attend on the day and at the place specified in the subpoena, or fails to remain in attendance until the inspector has excused him from further attendance;

(k) having been called under section 32, without sufficient cause (the onus of proof whereof shall rest upon him) –

(i) refuses to appear before the inspector;

     (ii) refuses to be sworn or to make affirmation as a witness after he has been directed to do so;

     (iii) refuses to answer, or fails to answer to the best of his knowledge and belief, any question put to him; or

     (iv) refuses to comply with a requirement to produce a book, document or thing specified in the subpoena or which he has with him;

(*l*) tampers with or discourages, threatens, deceives or in any way unduly influences any person with regard to evidence to be given or with regard to a book, document or thing to be produced by such a person before an inspector under section 32;

(*m*) prejudices, influences or anticipates the proceedings or findings of an inquiry under section 32 or 33;

(*n*) tampers with or misuses any safety equipment installed or provided to any person by an employer or user;

(*o*) fails to use any safety equipment at a workplace or in the course of his employment or in connection with the use of plant or machinery, which was provided to him by an employer or such a user;

(*p*) wilfully or recklessly does anything at a workplace or in connection with the use of plant or machinery which threatens the health or safety of any person,

shall be guilty of an offence and on conviction be liable to a fine not exceeding R50 000 or to imprisonment for a period not exceeding one year or to both such fine and such imprisonment.

(2) Any employer who does or omits to do an act, thereby causing any person to be injured at a workplace, or, in the case of a person employed by him, to be injured at any place in the course of his employment, or any user who does or omits to do an act in connection with the use of plant or machinery, thereby causing any person to be injured, shall be guilty of an offence if that employer or user, as the case may be, would in respect of that act or omission have been guilty of the offence of culpable homicide had that act or omission caused the death of the said person, irrespective of whether or not the injury could have led to the death of such person, and on conviction be liable to a fine not exceeding R100 000 or to imprisonment for a period not exceeding two years or to both such fine and such imprisonment.

(3) Whenever a person is convicted of an offence consisting of a failure to comply with a provision of this Act or of any direction or notice issued thereunder, the court convicting him may, in addition to any punishment imposed on him in respect of that offence, issue an order requiring him to comply with the said provision within a period determined by the court.

(4) Whenever an employer is convicted of an offence consisting of a contravention of a provision of section 23, the court convicting him shall inquire into and determine the amount which contrary to the said provision was deducted from the remuneration of the employee concerned or recovered from him and shall then act with respect to the said amount mutatis mutandis in accordance with sections

28 and 29 of the Basic Conditions of Employment Act, 1983 (Act No. 3 of 1983), as if such amount is an amount underpaid within the meaning of those sections.

2.12 CONSTRUCTION NEAR-MISSES

A near miss in construction is an accident on a construction site that has simply not yet resulted in an observable or visible physical consequence. All near misses should be investigated in the same way that accidents are. If treated properly, the collection, analysis and investigation of near misses will prevent the injury and damage outcomes of those events if remedial action is identified and applied. Targets set to reduce lost-time accidents will be achieved due to an underreporting culture. Dealing with near misses is injury prevention and not accident prevention. There are more 'no injury' accidents in construction than those that actually result in injury. There is still the problem of overcoming the reluctance or inertia of workers who may never report a near miss. Reasons for underreporting by workers include the following:

- The fear of blame
- The apparent triviality of the event measured by its outcome
- Embarrassment
- Suspicion that nothing will be done
- Perceived enormity of the task of filling in forms
- Apathy
- Lack of a workable system
- Personal fault.

Using the principle that if one does not ask one does not get, management must insist that all near misses are reported and fully investigated.

2.13 COST OF CONSTRUCTION INCIDENTS AND ACCIDENTS

The primary goal of injury and illness prevention according to Occupational Safety and Health Administration (OSHA) in the USA is to prevent workplace injuries, illnesses and deaths, the suffering these events cause workers, and the financial and emotional hardship they cause both workers and employers. Combined with inadequate medical aid or insurance, construction workplace injuries and illnesses can cause not only physical pain and suffering to the affected worker, but also loss of employment and wages, burdensome debt, inability to maintain a previous standard of living, loss of homeownership and even bankruptcy.

All unwanted construction incidents[4] and accidents cost money. Total costs are often difficult to measure, but there are always hidden expenses that must be identified and taken into account. Even a basic construction incident or accident costs a great deal.

---

4   Incidents include near misses.

Costs must be measured in terms of the effect on the individual and the community, as well as the financial loss.

It has been stated that if the true costs of construction injuries were well defined, management would be in a better position to make informed decisions concerning health and safety. Rather than addressing construction health and safety solely from an altruistic point of view, employers should also consider health and safety from an economic perspective.

SOCIAL IMPACT

All elements of society such as construction workers, families, employers, the economy and resources have probably been affected somehow by the occurrence of an accident at work. The economic impacts of construction incidents and accidents are, for example, a decrease in family income, a decrease in the standard of living, negative effects on education and schooling expenses, increase in debts and difficulty to pay bills, insurance policies and mortgage bonds. Victims and their dependents could suffer great pain, discomfort, hardship, sorrow and even psychological effects for many years following a disabling occupational disease or injury. The loss of earning power of someone who has contracted asbestosis or lost an arm will affect him or her in an unimaginable way.

FINANCIAL IMPACT

Figure 2.2 depicts the situation quite clearly, namely, that the portion of the iceberg below the surface is far greater than the exposed peak. This applies to the cost of a loss-producing incident or accident. The accumulative effects of all the construction incidents have a dramatic effect on the profits or bottom line of an organisation, as well as the national economy.

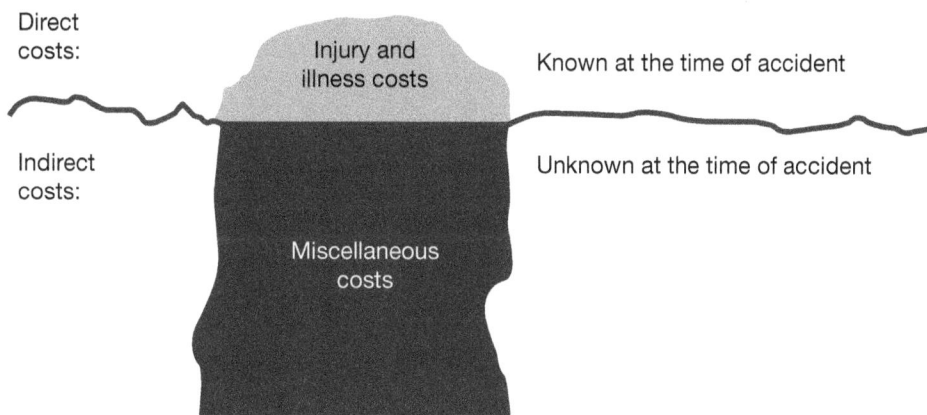

Figure 2.2: Iceberg depicting the relationship between direct and indirect costs of accidents

The costs of construction occupational injuries and illnesses can be divided into the following broad categories:

1. Direct costs include the costs of hospital, physician and allied health services, rehabilitation, nursing, home care, home health care, medical equipment, burial costs, insurance administrative costs for medical claims, payments for mental health treatment, police, fire, emergency transport, coroner/mortuary services and property damage.

2. Indirect costs refer to construction worker productivity losses that include wage losses and household production losses, administrative costs that include the cost of administering construction workers' compensation wage replacement programmes and sick leave, disruption of work schedules, administrative time for investigations and reports, training of replacement personnel, wages paid to the injured construction workers and others for time not worked, clean up and repair, adverse publicity, third-party liability claims, and construction plant and equipment damage.

3. Costs of health and safety programmes include salaries for health and safety, medical and clinical personnel, health and safety meetings, inspections of tools and equipment, orientation or induction sessions, workplace inspections, personal protective equipment, health programmes, and miscellaneous supplies and equipment.

4. Quality-of-life costs refer to value attributed to the pain and suffering that victims and their families experience as a result of injury, fatality or illness. Loss of quality-of-life costs can be six times greater than the construction project disruption costs.

The indirect costs of construction incidents and accidents are greater than their direct costs. Estimates of indirect costs as a proportion of direct costs have ranged from ratios that are 1:1 to more than 20:1, depending on the specific industry and methodology utilised. OSHA often uses the ratio from a study done at Stanford University of 1.1 for the most severe injuries to 4.5 for the least severe injuries times the actual calculable direct costs. Given that indirect costs of construction accidents always exceed the direct costs, it makes business sense to give greater attention to the indirect costs of construction worker injuries. Even where indirect and direct costs are similar, indirect costs balloon exponentially when liability litigation is instituted for a construction worker's injury. The following example illustrates how much an injury could potentially affect the output or bottom-line of a construction company trying to achieve a 3% profit margin on a construction project. If the company sustained a ZAR50 000 loss due to injury, illness or damage and still tries to make a 3% profit, the company theoretically must increase construction project income by an additional ZAR1 667 000.

The Health and Safety Executive in the United Kingdom suggest the following Accident Cost Calculator to determine and record the cost of incidents and accidents at work.

Table 2.3: Accident Cost Calculator

Category	Example	Time spent	Cost (£)
Dealing with incident (immediate action)	• First aid treatment • Taking injured person to hospital/home • Making the area safe • Putting out fires • Immediate staff downtime • Other		
Getting back to business	• Assessing/rescheduling work activities • Recovering work/production (including staff costs) • Cleaning up site and disposal of waste, equipment, products, etc • Bringing work up to standard such as product reworking time/cost • Repairing any damage/faults • Hiring or purchasing tools, equipment, plant and services • Other		
Business costs	• Salary costs of injured person while off work • Salary costs of replacement workers • Lost work time including people waiting to resume work, delays, reduced productivity, and effects on the productivity of other workers • Overtime costs • Recruitment costs for new workers • Contract penalties • Cancelled and/or lost orders • Other		
Action to safeguard future business	• Reassuring customers • Providing alternative sources of supply for customers • Other		
Sanctions and penalties	• Compensation claims payments • Attorneys' fees and legal expenses • Staff time dealing with legal cases • Fines and costs imposed due to criminal proceedings • Increase in insurance premiums • Other		
Other	• Own examples		
Total cost			

These are all important factors that must be considered in the estimation of the amount of monetary compensation. Each case will be different and the severity of the case will certainly affect the costs.

REVIEW QUESTIONS

1. What are the responsibilities of construction employers in terms of the Act?
2. Why are construction health and safety committees important in an organisation?
3. Describe how safe work or operating procedures and codes of practice contribute to construction health and safety on construction projects.
4. What is the role of the CEO of a construction company with respect to health and safety?
5. When is an inspector likely to issue a prohibition notice?
6. Why should near misses on construction sites be fully investigated?

REFERENCES

Bester, A & Haupt, TC. 2007. The role of the CEO in Construction Information Systems (IS) Performance: A pilot study. *Acta Structilia* 14(1): 56–80

Bird, F. 1974. *Management Guide to Loss Control.* Loganville, Georgia, US: Institute Press

Deacon, CH & Smallwood, J. 2017. *The Effect of the Integration of Design, Procurement, and Construction Relative to Health and Safety* (Doctoral dissertation, Nelson Mandela Metropolitan University)

Department of Labour. nd-a. *What Every Worker Should Know about Health and Safety in the Workplace* [online] Sizananitraining.co.za. Available at: http://www.sizananitraining.co.za/What%20every%20worker%20should%20know%20about%20Health%20and%20Safety%20in%20the%20Workplace.pdf.

Department of Labour. nd-b. *What Every Worker and Employer Should Know About Health and Safety in the Workplace.* [online] Labourguide.co.za. Available at: https://www.labourguide.co.za/health-and-safety/739-what-every-worker-and-employer-should-know-about

Department of Labour. 1993. Occupational Health and Safety Act 85 of 1993

Department of Labour. 1993. Compensation for Occupational Injuries and Diseases Act 103 of 1993

Department of Labour. 1995. Labour Relations Act 66 of 1995.

Department of Labour. 1997. Basic Conditions of Employment Act 75 of 1997

Department of Labour. 1998. Employment Equity Act 55 of 1998

Eppenburger, M & Haupt, TC. 2009. Construction worker injuries and costs – a comparative study of older and younger workers. *Occupational Health Southern Africa* 15(5): 6–13

Haupt, TC. 2021. *Management of Safety, Health and Environment in South Africa: A Handbook.* Newcastle upon Tyne: Cambridge Scholar Publishing

Haupt, T & Hefer, E. 2015a. Accident cost estimation relationship between direct and indirect costs. Proceedings of Association of Schools of Construction of Southern Africa (ASOCSA) 9th Built Environment Conference, 2-4 August, University of KwaZulu Natal, Durban, South Africa, pp 387–407

Haupt, T & Hefer, E. 2015b. How much is enough? A pilot study of the cost of construction health and safety. Proceedings of Association of Schools of Construction of Southern Africa (ASOCSA) 9th Built Environment Conference, 2-4 August, University of KwaZulu Natal, Durban, South Africa, pp 448–457

Haupt, T & Pillay, K. 2016. Investigating the true costs of construction accidents. *Journal of Engineering, Design and Technology* 14(2): 373–419

Hermanus, MA. 2007. Occupational health and safety in mining – status, new developments, and concerns. *Journal of the Southern African Institute of Mining and Metallurgy* 107(8): 531–538

Lewis, J & Thornbory, G. 2010. *Employment Law and Occupational Health*. Chichester, West Sussex: Wiley-Blackwell

Mpambane, S & Haupt, TC. 2007. An investigation into the effectiveness of the Inspectorate in South African home building industry. In Haupt, T. (ed) The Built Environment, Proceedings of the 2nd Built Environment Conference, Port Elizabeth, June 17-19, pp 187–196

Pillay, K & Haupt, TC. 2008a. The cost of construction accidents: An exploratory study. In Hinze, J, Bohner, S & Lew, J. (eds) Proceedings of the CIB W99 14th International Conference on Evolution of and Directions in Construction Safety and Health, pp 456–464

Pillay, K & Haupt, TC. 2008b. The cost of construction accidents: A pragmatic study. In Haupt, TC (ed) The Built Environment 3, Proceedings of the 3rd Built Environment Conference, Cape Town, July 6-8, pp 268–283

Scott, G. 2011. *An Assessment of Health and Safety Management in Selected Rural Hospitals.* [online] Dspace.nwu.ac.za. Available at: https://dspace.nwu.ac.za/bitstream/handle/10394/8437/Scott_GLS.pdf?

CHAPTER 3
STATUTORY REQUIREMENTS: THE REGULATIONS

3.1 REGULATIONS OF THE ACT

Section 43 of the Occupational Health and Safety Act 85 of 1993 (the Act) provides as follows with regard to regulations:

(1) The Minister may make regulations –

(a) as to any matter which in terms of this Act shall or may be prescribed;

(b) which in the opinion of the Minister are necessary or expedient in the interest of the health and safety of persons at work or the health and safety of persons in connection with the use of plant or machinery, or the protection of persons other than persons at work against risks to health and safety arising from or connected with the activities of persons at work, including regulations as to –

    (i) the planning, layout, construction, use, alteration, repair, maintenance or demolition of buildings;

    (ii) the design, manufacture, construction, installation, operation, use, handling, alteration, repair, maintenance or conveyance of plant, machinery or health and safety equipment;

    (iii) the training, safety equipment or facilities to be provided by employers or users, the persons to whom and the circumstances in which they are to be provided and the application thereof;

    (iv) the health or safety measures to be taken by employers or users;

    (v) the occupational hygiene measures to be taken by employers or users;

    (vi) any matter regarding the biological monitoring or medical surveillance of employees;

    (vii) the production, processing, use, handling, storage or transport of, and the exposure of employees and other persons to, hazardous articles, substances or organisms or potentially hazardous articles, substances or organisms, including specific limits, thresholds or indices of or for such exposure;

    (viii) the performance of work in hazardous or potentially hazardous conditions or circumstances;

    (ix) the emergency equipment and medicine to be held available by employers and users, the places where such equipment and medicine are to be held, the requirements with which such equipment and medicine shall comply, the inspection of such equipment and medicine, the application of first-aid and the qualifications which persons applying first-aid shall possess;

(x) the compilation by employers of health and safety directives in respect of a workplace, the matters to be dealt with in such directives and the manner in which such directives shall be brought to the attention of employees and other persons at such a workplace;

(xi) the registration of persons performing hazardous work or using or handling plant or machinery, the qualifications which such persons shall possess and the fees payable to the State in respect of such registration;

(xii) the accreditation, functions, duties and activities of approved inspection authorities;

(xiii) the consultations between an employer and employees on matters of health and safety;

(xiv) subject to section 36, the provision of information by an employer or user to employees or the public on any matter to which this Act relates;

(xv) the conditions under which any employer is prohibited from permitting any person to partake of food or to smoke on or in any premises where a specified activity is carried out;

(xvi) the conditions under which the manufacture of explosives and activities incidental thereto may take place;

(c) as to the preventive and protective measures for major hazard installations with a view to the protection of employees and the public against the risk of major incidents;

(d) as to the registration of premises where employees perform any work or where plant or machinery is used and the fee payable to the State in respect of such registration;

(e) whereby provision is made for the continuation of any registration under this Act;

(f) as to the registration of plant and machinery and the fee payable to the State in respect of such registration;

(g) as to the establishment of one or more committees for the administration of a provision of the regulations, the constitution of such committees, the functions of such committees, the procedure to be followed at meetings of such committees, the allowances which may be paid to members of such committees from money appropriated by Parliament for such purpose and the person by whom such allowances shall be fixed;

(h) prescribing the records to be kept and the returns to be rendered by employers and users and the person or persons to whom such returns shall be rendered;

(i) as to the designation and functions of health and safety representatives and health and safety committees and the training of health and safety representatives;

(j) as to the activities of self-employed persons; and

(k) as to any other matter the regulation of which is in the opinion of the Minister necessary or desirable for the effective carrying out of the provisions of this Act.

(2) No regulation shall be made by the Minister except after consultation with the Council, and no regulation relating to State income or expenditure or to any health matter shall be made by the Minister except after consultation also with the Minister of State Expenditure and the Minister for National Health and Welfare, respectively.

(3) In making regulations the Minister may apply any method of differentiation that he may deem advisable: Provided that no differentiation on the basis of race or colour shall be made.

(4) A regulation may in respect of any contravention thereof or failure to comply therewith prescribe a penalty of a fine, or imprisonment for a period not exceeding 12 months, and, in the case of a continuous offence, not exceeding an additional fine of R200 or additional imprisonment of one day for each day on which the offence continues: Provided that the period of such additional imprisonment shall not exceed 90 days.

(5) A regulation made under section 35 of the Machinery and Occupational Safety Act, 1983 (Act No. 6 of 1983), which was in force immediately prior to the commencement of this Act and which could have been made under this section, shall be deemed to have been made under this section.

The regulations in the Act are identified by their titles (*not* by numbers). All the regulations are available online.

- Asbestos Regulations, 2001
- Certificate of Competency Regulations, 1990
- Construction Regulations, 2014
- Diving Regulations, 2009
- Driven Machinery Regulations, 1988
- Electrical Installation Regulations, 2009
- Electrical Machinery Regulations, 1988 (most people use portable electrical tools)
- Environmental Regulations for Workplaces, 1987 (re lighting, ventilation, noise protection, and fire safety and precautions in the work environment)
- Explosives Regulations, 2003
- Facilities Regulations, 1990 (workers require seating, drinking water and sanitary facilities)
- General Administration Regulations, 2003 (management's administrative procedures with regard to health and safety in the workplace)
- General Machinery Regulations, 1988 (ensuring health and safety with regard to driven machinery)
- General Safety Regulations (refers to PPE, first aid facilities, use of ladders, limited access to high risk areas, welding equipment, stacking and storage, and eviction of intoxicated persons)

- Hazardous Biological Agents Regulations, 2001
- Hazardous Chemical Substances Regulations, 1995
- Incorporation of Safety Standards into Electrical Installation Regulations, 2009
- Lead Regulations, 2001 (eg in welding or use of lead-based paint)
- Lift, Escalator and Passenger Conveyor Regulations, 1994 (buildings with lifts)
- Major Hazard Installation Regulations (eg hazardous chemical stores)
- National Code of Practice: Evaluation of Training Providers for Lifting Machine Operators (eg maintenance workshops)
- Noise-induced Hearing Loss Regulations, 2003 (eg noise zones such as workshops and construction sites)
- Pressure Equipment Regulations, 2009 (eg fire extinguishers, diving and gas cylinders, and compressors)
- Regulations for the Integration of the Occupational Health and Safety Act, 1995 (refers to integration of Labour Laws)

3.2 CONSTRUCTION REGULATIONS

Increasing awareness relative to the role of H&S in overall construction project performance and the inclusion of H&S as a project performance measure in contrast to the traditional focus on time, cost and quality has engendered increasing focus on H&S by a range of construction stakeholders. The number of large-scale construction accidents in South Africa and the consequential media coverage of these has further raised the level of awareness. The Construction Regulations 2014 (CR) were promulgated by the Minister of Labour under section 43 of the Act on 7 February 2014 as a major revision about 11 years after they were first introduced on 18 July 2003. The revisions were necessary due to gaps found in the 2003 CR and the continuing poor performance of the construction sector. Additionally, the CR require a range of interventions by all participants in the construction process from clients and designers to workers themselves in the effort to improve the overall poor H&S performance of the sector.

Generally, the provisions of the CR herald a welcome and overdue departure, in particular from previous approaches to the management of construction safety and health. Previously contractors were held solely responsible for the safety and health of their workforce. The CR represent a major paradigm shift in that the responsibility for safety and health is redistributed away from contractors to include all participants in the construction process, with particular emphasis on the pivotal role of clients, rightly redistributing the responsibility for construction health and safety to include all participants in the construction process throughout the entire life cycle of a project. Furthermore, the CR recognise that the construction industry is client driven from project inception through to the eventual demolition or deconstruction of a facility or structure. They also recognise the pivotal role of clients in positively influencing the health and safety performance of the rest of the project team which includes contractors. This is necessary in an industry that is largely client driven.

Rather than setting and insisting on compliance with myriads of standards, the CR are performance driven. Instead, the emphasis is on hazard identification and risk assessment prior to the execution of construction projects and activities. Furthermore, all parties are encouraged to eliminate risks at source, reduce exposure to risks, or protect against the consequences of unavoidable risks. For the CR to be optimally effective it will take a broad-based commitment from all stakeholders – from client to worker on site – to conduct themselves in the spirit of the regulations rather than the letter.

In particular, clients are required to provide the principal contractor (PC) with an H&S specification, and ensure that PCs have made adequate allowance for H&S. Designers are required to provide the client with all relevant information about the design, which will affect the pricing of the works, inform the contractor of any known or anticipated dangers or hazards, provide the contractor with a geo-science technical report and the methods and sequence of construction, and modify the design where dangerous procedures would be necessary, or substitute hazardous materials. For the CR to be effective demands a major industry paradigm shift on the part of all construction process stakeholders – government included – from one of compliance with legislation, and more often than not minimum compliance, to one of changing the way construction is done in pursuit of overall better practice. Arguably, there would have been no need for the CR had the industry had its house in order. In fact, they should be seen as an indictment of the poor health and safety performance of the industry. The ongoing spate of fatalities and accidents on construction sites throughout South Africa reinforce the need for the CR.

Some of the features of the CR include the following:
- Key definitions
- Involvement of client at project inception
- Written application for a construction works permit which will be issued to the client
- Written notification of construction work by contractor
- Baseline risk assessment by client
- Site specific construction health and safety specification by client
- Inclusion of construction health and safety specification in tender documents
- Financial provision for construction health and safety in tenders
- All appointees in the construction team to have necessary competences and resources
- Contractor project and site-specific construction health and safety plan which is sufficiently documented and coherent, based on the documented health and safety specification of the client
- Risk assessments as part of the construction health and safety plan
- Registered and in good standing with compensation fund or compensation insurer by contractor
- Health and safety audits and document verification
- Consolidated health and safety file prepared by contractor to be handed over to the client upon completion of the project for post-construction use by the client

- Appointment by client of a health and safety agent where a construction work permit is required
- Specific designer construction health and safety responsibilities
- Authority to stop work by contractor
- Design of temporary works
- Valid medical certificates of fitness for all employees of contractors
- Mandatory site and project-specific health and safety induction or orientation programme developed and implemented by contractor
- Ergonomic hazards to be addressed in health and safety plan
- Mandatory fall protection plan
- Prevention of the uncontrolled collapse of any new or existing structure or any part thereof, which may become unstable or is in a temporary state of weakness or instability due to the carrying out of construction work
- Inspections of excavations
- Detailed structural engineering survey and preparation of method statement before demolition
- Management of the presence of asbestos and lead
- Preparation of a method statement prior to the use of explosives
- A certificate of system design must be issued by a professional engineer, certificated engineer or a professional technologist for the use of a suspended platform system as well as operational compliance plan
- Risk assessment and method statement required for use of cranes
- Establishment of a construction health and safety technical committee.

3.3 GENERAL ADMINISTRATIVE REGULATIONS 2003

The General Administrative Regulations (GAR) deal with various administrative issues regarding health and safety in the construction workplace, and provide for construction health and safety committees, negotiations and consultations before designation of construction health and safety representatives, reporting of incidents and occupational diseases, recording and investigation of incidents, and other related matters. They require that:
- employers ensure that a copy of the Act is available on the construction site, in the plant, store or workshop
- any person has the right to access the Act
- workers report incidents to supervisors and management.

3.4 EXAMPLES OF FACILITY REGULATIONS

The following are examples of requirements for facilities on the construction site, in the plant, store or workshop:
- **Sanitary facilities** must have signs for males and females; be clean and hygienic;

have water, toilet paper, soap or alcohol-based hand sanitiser and disposable paper towel or blower; and be in a working condition.

- **Safekeeping facilities** must be kept sanitised; available to construction workers to keep personal belongings in them; and workers must use them.
- **Change rooms** must have signs for males, females and persons with disabilities; not be connected to areas which can contaminate them and must be kept clean and hygienic and regularly sanitised; lockers, chairs and benches must be a minimum of 1.5 m apart to ensure social distancing.
- **Dining rooms** where provided must be situated outside the work area, and social distancing of a minimum of 1.5 m between tables and chairs, hand sanitising stations, wearing of face masks and limited numbers of workers using them at one time must be enforced.
- **Drinking water** must be provided where existing water is unfit for human consumption and proper informative signage is to be provided.
- **Proper seating** that does not cause back pains or poor posture must be provided to construction workers who can do their work in a sitting position, while ensuring social distancing of at least 1.5 m between their work stations.

3.5 EXAMPLES OF ENVIRONMENTAL REGULATIONS FOR WORKPLACES

The following are examples of requirements for environmental regulations on the construction site, in the plant, store or workshop:

Temperature
In cold conditions the correct protective clothing is required for construction workers working outside in winter. When working with vibrating tools in winter workers must wear padded gloves and hearing protection. In temperatures below 0°C workers must be certified medically fit and wear the right protective clothing.

On the other hand, in hot conditions, particularly where hard manual labour is involved, construction workers must be certified medically fit, must be provided with enough water, and first aid must be available for treatment of construction workers suffering from heat stroke.

Housekeeping
A storage facility on the construction site, in the plant, store or workshop should be provided for whatever is not in use, whether they be materials, tools, plant or equipment.

Fire precautions
Firefighting equipment on the construction site, in the plant, store or workshop must be strategically placed, sign-posted, unobstructed, regularly serviced (check service dates) and in a good condition. All construction workers must know how to use them.

3.6 NOISE INDUCED HEARING LOSS REGULATIONS

The condition and positioning of proper hearing protection signs are important. All construction workers and visitors entering and remaining in noisy areas on the construction site, in the plant, store or workshop must wear the required hearing protection regardless of how long they remain in a noise zone, and the hearing protection equipment must be stored in the facility provided when not in use.

Figure 3.1: An example of the level of noise generated by demolition operations using mechanical demolition equipment in Ankara, Turkey

3.7 GENERAL SAFETY REGULATIONS

The following General Safety Regulations (GSRs) must be implemented on the construction site, in the plant, store or workshop:

Personal protective equipment
Signs must be posted where personal protective equipment (PPE) must be worn. All construction workers must know how to use or wear it and properly maintain it. PPE, which is a measure of last resort, must always be in a good and clean condition.

First aid

Construction workers must know who the first aiders are. Their names must be on the first aid box and they must be readily available to attend to anyone needing their services. The contents of the first aid box should comply with regulation 3(9). The content must be checked regularly. It is advisable for the first aid box to be sealed to prevent abuse and pilfering of the contents.

Flammable liquid store

Any flammable liquid store on a construction site must have a sign clearly demarcating it as a 'FLAMMABLE LIQUID STORE'. It needs to be properly ventilated to the outside atmosphere with a fan that must be switched on. The store may not contain more than 110% of the liquid to be stored in it. The correct type of firefighting equipment must be provided and strategically placed.

Stacking

All stacking of materials to be used on the construction site must be carried out under supervision of an experienced construction worker or supervisor. Any pallets used for stacking must not be broken and whatever is stacked on them must be stable. Access to and from any stacks must be safe.

Welding and flame cutting

All welding and flame cutting activities on the construction site must be screened off. All electric leads must be insulated. The correct protective equipment must always be worn such as protection of eyes, hands, legs, feet and respirators when welding is done, particularly in confined spaces. No welding must be done on closed containers that are explosive and could ignite. When welding inside metal vessels always ensure there is a stand-by person in case of an emergency.

Intoxication

No worker is allowed to be intoxicated on the construction site, in the plant, store or workshop, or partake of or offer intoxicating substances, including cannabis, to other persons.

Ladders

Ladders must have non-skid devices and be in a good condition. They must only be used if the following are in place:
- Hooks at upper end to ensure stability, or
- Held by another construction worker depending on the height at which the work is to be performed, or
- Lashed or secured by any other means to the structure being worked on
- Ladders are NOT allowed to be lashed together.

Any ladders leaning against an object or structure:
- are not allowed to be longer than 9 m or they will become unstable
- must not be painted and must be inspected regularly if they are wooden.

3.8 OTHER RELATED LEGISLATION

3.8.1 BASIC CONDITIONS OF EMPLOYMENT ACT

The Basic Conditions of Employment Act (BCEA) covers issues such as working time or hours, and maximum permitted hours of work on the construction site and in the plant, store or workshop. Taking working time schedules into consideration, when providing and maintaining a safe and healthy working environment, arrangements should be considered for the special needs of construction workers, for example:
- Perform critical tasks safely without risk to health
- Health assessment and counselling
- Services should be available at night if work activities occur at night
- Daily and weekly rest periods
- Annual leave
- Protection of children
- Protection against discrimination
- Protection of women
- Protection before and after birth.

3.8.2 LABOUR RELATIONS ACT, AS AMENDED

The Labour Relations Act (LRA) deals with, for example:
- Protection against victimisation on the construction site, in the plant, store or workshop
- Protected disclosures
- Dismissal for non-compliance with construction health and safety rules
- Dismissal for endangering the construction worker and others on the construction site, in the plant, store or workshop
- Work-related injuries
- Training
- HIV and AIDS and incapacity.

3.8.3 EMPLOYMENT EQUITY ACT

The Employment Equity Act (EEA) covers, for example, the following:
- Elimination of unfair discrimination in recruitment processes and on the construction site, in the plant, store or workshop
- Implementation of affirmative action to redress inequalities of the past
- People with disabilities

- Prohibition of medical testing as a condition for employment
- Prohibition of HIV testing without permission of the construction worker.

3.8.4 SKILLS DEVELOPMENT ACT

The Skills Development Act (SDA) as it applies to construction deals with, inter alia, construction H&S Training and claiming levies.

REVIEW QUESTIONS

1. What are some of the key elements of the Construction Regulations 2014?
2. How would you deal with the use of intoxicating substances on your construction sites?
3. Develop COVID-19 protocols for your construction organisation and show the practical steps to be taken in all areas of the construction site, plant, store or workshop.
4. Describe how you would manage the use of personal protective equipment (PPE) on the construction site, in the plant, store or workshop.
5. If you have a dining room on the construction site, in the plant, store or workshop or want to provide such a facility, what are some of the H&S considerations you would need to consider?

REFERENCES

Department of Environmental Affairs. 1965. Atmospheric Pollution Prevention Act 45 of 1965

Department of Environmental Affairs. 1989. Environmental Conservation Act 73 of 1989

Department of Environmental Affairs. 1998. National Environmental Management Act 107 of 1998

Department of Environmental Affairs and Tourism. 2004. Environmental Management Plans, Integrated Environmental Management, Information Series 12, Department of Environmental Affairs and Tourism (DEAT), Pretoria

Department of Labour. 1997. Basic Conditions of Employment Act 75 of 1997

Department of Labour. 1998. Employment Equity Act 55 of 1998

Department of Labour. 1998. Skills Development Act 97 of 1998

Department of Labour. 1995. Labour Relations Act 66 of 1995

Department of Labour. 2001. Noise Induced Hearing Loss Regulations, 2001. GNR 681, G 22499

Department of Labour. 2003. General Administration Regulations, 2003. GoN R929, G 25129

Department of Labour. 2003. General Safety Regulations, 2003. GoN R1010, G 25207

Department of Labour. 2004. Facilities Regulations, 2004. GNR 924

Department of Labour. 2017. Construction Regulations, 2014. *Government Gazette* 40883

Haupt, TC. 2019a. Challenges of construction in South Africa – Research opportunities to challenges of construction. Presentation to staff, Massey University, New Zealand, 27 September 2019

Haupt, TC. 2019b. The state of construction health and safety in South Africa. Presentation to final year engineering students and staff, Curtin University, Australia, delivered 10 September 2019

Haupt, TC. 2020. Compliance or better practice: Health and safety in uncertain times. Webinar delivered for University Extended Learning, October 13, 2020

Haupt, TC. 2021. *Management of Safety, Health and Environment in South Africa: A Handbook*. Newcastle upon Tyne: Cambridge Scholar Publishing

Raliile, MT & Haupt, TC. 2019. Analysis of recent construction regulations changes and their impact on the quality of life of construction workers. Proceedings of the 1st Association of Researchers in Construction Safety, Health and Wellbeing (ARCOSH) Conference, Cape Town, South Africa, pp 140–146

Raliile, M & Haupt, TC. 2020a. Involvement in the implementation of health and safety policies on construction sites. Joint CIB W099 & TG59: International Good Health, Wellbeing & Decent Work Conference Glasgow

Raliile, M & Haupt, TC. 2020b. The study on knowledge, attitudes and commitment of managers within construction firms towards recent construction health and safety legislation changes. 2nd Association of Researchers in Construction Safety, Health and Wellbeing Conference (ARCOSH), Cape Town, South Africa

Windapo, A & Oladapo, AA. 2012. Determinants of construction firms' compliance with health and safety regulations in South Africa. In Procs of 28th Annual ARCOM Conference, 3-5 September 2012, Edinburgh, UK (Vol 2, pp 433–444). Association of Researchers in Construction Management (ARCOM)

CHAPTER 4
RESPONSIBILITIES OF
CONSTRUCTION PROCESS
STAKEHOLDERS

4.1 INTRODUCTION

While it should not be the only reason to implement construction health and safety, the South African legislative and regulatory framework for health and safety on construction projects and sites places responsibilities on all stakeholders involved in the construction process, from clients to workers on site. These responsibilities are required to be met during all the phases of a construction project from its inception to its final demise at the end of its lifespan through demolition and deconstruction. Each of the various stakeholders can influence the overall construction health and safety performance on a project to varying degrees if they have the will to do so, irrespective of the legal requirements to do so, which are minimum requirements at best.

4.2 CLIENTS

In construction a client is any person for whom construction work is being performed: in other words, every person or entity that enters into a contract to have construction work executed on their behalf. Therefore, clients initiate construction projects and advise their project design teams. They dictate constraining factors, such as the project schedule and budget, which can have an impact on construction health and safety. They also decide on the project procurement strategy, performance monitoring mechanisms and contractual requirements for their projects. An appropriate use of procurement procedures and contract documentation has the potential to raise the standard of H&S on individual projects. Therefore, procurement is a direct way for clients to make a real difference to H&S in their area of influence.

Given this strategic role, clients are well placed to both directly and indirectly influence the construction health and safety improvement effort on their projects. The incentive to contribute includes the following benefits:
- Reduced cost of construction because of increased productivity on site
- Avoidance of negative publicity because of accidents and fatalities
- No disruption to processes (alterations or extensions) due to, for example, delays caused by accidents, investigations and damage to property.

In terms of construction projects in South Africa, the client has specific roles and responsibilities which include the following:

- Apply in writing for a construction work permit to perform construction work if the intended construction work:
  - will exceed 180 days;
  - will involve more than 1 800 person days of construction work; or
  - the works contract is of a value equal to or exceeding 13 million rand; or
  - involves Construction Industry Development Board (CIDB) grading level 6.

  A copy of the construction work permit must be kept in the occupational health and safety file for inspection by an inspector, the client, the authorised agent of the client or a worker.
- Prepare a baseline hazard identification and risk assessment (HIRA) for the intended construction work project.
- Prepare a suitable, sufficiently documented and coherent site-specific health and safety specification for the intended construction work based on the baseline risk assessment.
- Provide the designer with the health and safety specification.
- Ensure that the designer takes the provisions of the prepared health and safety specification into consideration when developing the design of the construction project so that the designer can mitigate in the actual design the exposure to the significant hazards identified in the baseline risk assessment by considering the following hierarchy of controls:
  - elimination of the hazards themselves
  - substitution of materials, construction systems and methods
  - engineering controls
  - administrative controls
  - personal protective equipment as a measure of last resort where additional mitigating interventions are not possible.
- Include the health and safety specification in the tender documents as a means of the client influencing the project procurement strategy, performance monitoring mechanisms and contractual requirements of their projects.
- Ensure that potential principal contractors submitting tenders have made adequate financial provision in their tenders for the cost of health and safety measures to mitigate the exposure to the significant hazards identified in the baseline HIRA.
- Ease the process of checking that adequate provision for H&S has been made by listing items that are necessary to meet the requirements of the client and which can be separately priced as a prime cost item in the bill of quantities, provisional sum, or in whatever other pricing mechanism is used.
- Ensure that the principal contractor to be appointed has the necessary competencies and resources to carry out the construction work safely and without any threat to the health and well-being of everyone involved in the project, including the public at large.

- The potential principal contractor must be prequalified as part of the tender adjudication process prior to making the appointment.
- Take reasonable steps to ensure cooperation between all contractors appointed by the client to enable each of those contractors to carry out the construction work safely without any threat to the health and well-being of everyone involved in the project, including other contractors and the public at large.
- Ensure before any construction-related work commences on a site that every principal contractor is registered and in good standing with the Compensation Fund or with a licensed compensation insurer as contemplated in the Compensation for Occupational Injuries and Diseases Act 130 of 1993.
- Formally appoint every principal contractor in writing for the project or part thereof on the construction site.
- Discuss and negotiate with the principal contractor, either directly or by using a competent and resourced H&S practitioner or consultant, the contents of the project- and site-specific health and safety plan produced by the principal contractor and must thereafter finally approve that plan for implementation.
- Ensure that a copy of the approved health and safety plan of the principal contractor is available on request to a worker, inspector or contractor.
- Take reasonable steps to ensure that the approved health and safety plan of each contractor appointed to work on the project is implemented and maintained.
- Ensure that periodic health and safety audits and document verification are conducted at intervals mutually agreed upon between the principal contractor and any contractor, but at least once every 30 days during the project construction period.
- Ensure that a copy of the health and safety audit report is provided to the principal contractor within seven days after the audit was conducted.
- Stop any contractor from executing any construction activity that poses a threat to the health and safety of anyone involved in the project, including other contractors and the public at large.
- Where changes are brought about to the design or construction work through instructions from the designer or variation orders, ensure that sufficient health and safety information and appropriate resources are available to the principal contractor to execute the additional work safely and without any threat to the health and well-being of everyone involved in the project, including other contractors and the public at large.
- Ensure that the principal contractor makes adequate financial provision in their pricing of the variations to the project for the cost of any additional health and safety measures that may be needed, especially if the additional work is out of sequence; and that the project schedule or programme is adjusted to take these variations into account.
- Ensure that the approved health and safety plan is amended as required to consider the mitigation of any hazards as a result of the additional work to be done.
- Ensure that the health and safety file is kept and maintained by the principal contractor.

- It is critically important to note that where a client requires additional work to be performed as a result of a design change or an error in construction due to the actions of the client, the client must ensure that sufficient health and safety information and appropriate additional resources are available to execute the required additional work safely and without any threat to the health and well-being of everyone involved in the project, including other contractors and the public at large.
- Where a fatality or permanent disabling injury occurs on a construction site, the client must ensure that the contractor provides the provincial director with a report in accordance with regulations 8 and 9 of the General Administrative Regulations, 2013, and that the report includes the measures that the contractor intends to implement to ensure a safe construction site as far as is reasonably practicable
- Where more than one principal contractor is appointed, the client must take reasonable steps to ensure cooperation between all principal contractors and contractors.
- Where a construction work permit is required, the client must, without derogating from his or her health and safety responsibilities or liabilities, appoint a competent person in writing as a construction H&S agent to act as his or her representative, and where such an appointment is made, all the duties of a client apply as far as is reasonably practicable to the appointed construction H&S agent.
- Where notification of construction work is required, the client may, without derogating from his or her health and safety responsibilities or liabilities, appoint a competent person in writing as a construction H&S agent to act as his or her representative, and where such an appointment is made, all the duties of a client apply as far as is reasonably practicable to the appointed construction H&S agent, provided that, where the question arises as to whether a construction H&S agent is necessary, the decision of an inspector is decisive.
- Prior to appointing a construction H&S agent, the client must prequalify the potential agent to ensure that they have the necessary competencies and are adequately resourced to carry out their duties on the project.

Note that the overall responsibility for construction health and safety is a non-delegable one and the principle of vicarious liability applies, in that the client still remains liable for construction health and safety of his project.

4.3 CLIENT-APPOINTED CONSTRUCTION HEALTH AND SAFETY AGENTS

Given the likelihood that clients are not knowledgeable enough to be able to fulfil the important non-delegable functions expected from them in terms of ensuring optimum construction health and safety performance on their construction projects, they must consider the formal appointment of a construction H&S agent who is competent and is adequately resourced to perform the duties of the client instead. Where such an appointment is made by the client, the client-appointed construction health and safety agent (CHSA) who has to be registered with the South African Council for the Project and Construction Management Professions (SACPCMP), must manage all aspects of the health and safety on the construction project for the client as if he were the client himself.

The CHSA cannot be a novice given the level of responsibility placed on him to perform all the H&S duties and functions of the client on his behalf. Therefore, the CHSA should have at least five years' practical experience in the construction industry, and should have been assessed by the SACPCMP as competent to provide professional health and safety services across the six construction H&S project stages as shown in Table 4.1.

Table 4.1: SACPCMP construction health and safety project stages

Stage 1: Project Initiation and Briefing
Stage 2: Concept and Feasibility
Stage 3: Design and Development
Stage 4: Tender Documentation and Procurement
Stage 5: Construction Documentation and Management
Stage 6: Project Close-out

The level of competence and knowledge required as a minimum, as well as accountability are considerable as the CHSA represents the client as its agent in matters relating to construction H&S through the entire project life cycle. The level of H&S competence and knowledge needs to enable the CHSA to relate to the complexity of the project design.

The SACPCMP requires knowledge of and competence in the following nine knowledge areas:

1. Procurement management
2. Cost management
3. Hazard identification management
4. Risk management
5. Accident or incident investigation management
6. Legislation and regulations management
7. Health, hygiene and environmental management
8. Communication management
9. Emergency preparedness management.

The CHSA must be able to design systems and develop appropriate responses to the H&S hazards that will likely be experienced on the project, compile relevant documents and reports for project H&S, implement H&S management systems as applicable, and manage all aspects of construction H&S across all the stages of the project. The CHSA should ensure that the integration of construction H&S extends to all project planning, particularly because the greatest effect on a project are those interventions that are taken early in the project life cycle.

In order to be registered with the SACPCMP, the CHSA has to, inter alia, do the following:
- Demonstrate that they have recognised technical qualifications in the construction H&S environment
- Demonstrate that they have at least five years' experience in the construction industry, particularly with respect to construction H&S management and implementation
- Demonstrate their knowledge, skills and experience to the SACPCMP by means of completing the requisite testing administered by the SACPCMP
- Be assessed by the SACPCMP as able to act suitably as a CHSA following a psychometric examination
- Be certified as competent in the transfer of construction H&S skills and knowledge
- Undergo an interview with the SACPCMP Health and Safety Agent Registration Committee.

4.4 DESIGNERS

In the South African construction industry, a designer is defined as:
- A person who prepares a design, for example, an architect, architectural technologist, architectural engineer or draughtsperson
- A person who checks and approves a design, for example, in the municipal office where plans are submitted for approval prior to construction
- A person who arranges for any person at work under his control to prepare a design, for example, a senior architect or engineer
- An architect or engineer contributing to or having overall responsibility for the design
- Building services engineer designing details for fixed plant, for example, air-conditioning, sanitary, water and drainage services
- Surveyor specifying articles or drawing up specifications, for example, a quantity, construction or building surveyor
- Contractor carrying out design work as part of a design and building project
- Temporary works engineer designing formwork and false work, for example, designing scaffolding framework for a building
- Interior designer, shopfitter and landscape architect.

In South Africa, designers are also legally and morally liable, in terms of section 10 of the OHS Act, for the impact on construction H&S of their design, which occurs upstream from the actual construction activities on site. The Construction Regulations specifically address the involvement of designers in construction H&S as stakeholders in the construction industry and process. Similarly, in the United Kingdom, the Construction (Design and Management) (CDM) Regulations (CDMR) link clients and designers fostering the integration of the design process with the construction process, reducing the fragmentation in the industry which has created adversarial relationships between the various stakeholders in the construction process. As design influences construction, there is a domino, or 'knock on' effect and H&S risks during the construction phase could well be increased if no consideration is given to either aspect.

Designers dictate the various aspects relating to the components of a facility through their design documents. Because designers give professional advice to their clients during the concept and general design phases of a construction project, they are well positioned to directly influence the impact of their designs on construction H&S when they select the structural frame develop sections and details through elements of the structure, identify positions of services to the structure taking into account ergonomic considerations, and specify materials, finishes and processes. Designers have the opportunity to consider H&S during the coordination of their designs and during constructability or buildability reviews of their designs as they are developed. Designers also set the parameters for civil, electrical, interior, landscape, mechanical and structural designers who use their designs as the basis or template for the design and layout of their own specialist installations.

There are many opportunities for designers to refer to construction H&S throughout the project delivery process. Many of the common health and safety problems encountered during the construction and operation phases could be avoided if due consideration and effort were invested during the project brief and design phases. Arguably the greatest opportunity to implement technological controls for H&S risk, namely those that eliminate or reduce a risk by adopting engineering solutions, is present during the early project planning and design stages. These include, for example, during pre-tender site visits, upon initial site handover, during various visits to site to conduct inspections and attend project progress meetings. With respect to their involvement in the choice of the structural frame of their design they have several options, for example, when opting for loadbearing masonry the mass of the bricks, extent of the effects of manual handling on the body when bending and twisting to lay the bricks, and the need for proper housekeeping when the bricks and related materials are stored on site. On the other hand, selecting a timber frame introduces other options such as prefabrication and offsite manufacture of components that are lighter than masonry units, less on site-storage and less manual handling as the timber components can be hoisted mechanically into place using a small crane or lifting device. The latter option invariably facilitates better H&S on site. Other considerations when selecting a structural frame include its impact on constructability, work method and work postures of workers, type of materials, plant and equipment required, means of access to work areas and design and provision of mechanical, electrical and plumbing services (MEP) which will have ergonomic impacts on how they are to be installed in their final positions including working space and fixing methods. Designers need therefore to consider the proper coordination of all work, layering to avoid clashes of piping and ducting runs, and sectional areas of vertical ducts to reduce the need to work in awkward positions.

When materials are specified for the structural frame, designers should take cognisance of the mass of particular materials per unit, linear metre, square area or volume. Further, whether the materials could be toxic, have rough surfaces with sharp edges, or be large or small, which will affect their manual handling. They should consider whether the shape of the structure is regular or irregular on the plan, which might require formwork or support work that must be custom designed. Other aspects that designers should consider include the following:

- Need to work at heights on external finishes to the building facades
- Pitch of roofs where steep or high pitches present as fall hazards
- Any oversailing or cantilevered sections that make access to their soffit challenging because of the need to work overhead
- Depths of excavations for drainage and sewerage lines to connect to municipal services, which might require special shoring and support work to prevent collapses.

Therefore, designers can encourage better H&S on construction sites by making strategic choices to mitigate exposures to hazards that could be eliminated or reduced by considering the hierarchy of control. Some specific considerations include the following:

- Considering H&S throughout all stages of design, namely during the client brief, concept design, detailed design, working drawings, sections and details
- Promoting where possible offsite prefabrication, pre-assembly and precasting
- Attempting to reduce work at elevated heights
- Specifying self-finished materials
- Engendering of mechanisation where possible
- Selecting appropriate procurement or project delivery systems
- Providing for H&S in project contract documentation
- Referring to H&S during pre-tender, pre-contract and contract phases of construction, and during commissioning and maintenance phases of projects, and recycling, reuse and deconstruction.

In summary, designing for H&S is the amount of consideration by designers of construction site H&S in the preparation of plans, specifications and contract documents to be used in their construction projects. It is evident that many designers do not accept that H&S and design are complementary or apply to them. Other designers only do what they have been instructed to do or what was specified in their contract with the client, and therefore are not going to be paid to design for H&S. Many designers do not believe that they have any impact on construction H&S, and that H&S is the sole responsibility of the contractor. These attitudes, perceptions and beliefs arise from designers not necessarily spending enough time on construction sites during the construction phase of their projects.

In terms of South African regulations, the designer of a structure must specifically do the following:

- Ensure that the applicable health and safety measures are complied with in the design.

- Take into consideration the health and safety specification submitted by the client.
- Before the contract is put out to tender, make available in a report to the client the following:
  - all relevant health and safety information about the design of the relevant structure that may affect the pricing of the construction work
  - the geotechnical-science aspects, where appropriate
  - the loading that the structure is designed to withstand.
- Inform the client in writing of any known or anticipated dangers or hazards relating to the construction work, and make available all relevant information required for the safe execution of the work upon being designed or when the design is subsequently altered.
- Refrain from including anything in the design of the structure necessitating the use of dangerous procedures or materials hazardous to the health and safety of persons that can be avoided by modifying the design or by substituting materials.
- Consider the hazards relating to any subsequent maintenance of the relevant structure and make provision in the design for that work to be performed to minimise the risk.
- When mandated by the client to do so, carry out the necessary inspections at appropriate stages to verify that the construction of the relevant structure is carried out in accordance with the design.
- If the designer is not so mandated, the client's appointed H&S agent is responsible to carry out such inspections.
- When mandated, stop any contractor from executing any construction work which is not in accordance with the relevant health and safety aspects of the design.
- If the designer is not so mandated, the client's appointed H&S agent must stop that contractor from executing that construction work.
- When mandated, in his or her final inspection of the completed structure in accordance with the National Building Regulations, include the health and safety aspects of the structure as far as reasonably practicable, declare the structure safe for use with no health threats, and issue a completion certificate to the client and a copy of the completion certificate to the contractor.
- During the design stage, take cognisance of ergonomic design principles to minimise ergonomic related hazards in all phases of the life cycle of a structure.

With respect to temporary works, the designer must ensure that:
- all temporary works are adequately designed so that they will be capable of supporting all anticipated vertical and lateral loads that may be applied
- the designs of temporary works are done with close reference to the structural design drawings issued by the contractor, and in the event of any uncertainty consult the contractor
- all drawings and calculations pertaining to the design of temporary works are kept at the office of the temporary works designer and are made available on request by an inspector

- the loads caused by the temporary works and any imposed loads are clearly indicated in the design.

4.5 CONTRACTORS IN GENERAL

For too long contractors were held solely responsible for construction H&S and other stakeholders in the construction process supported this view and as a result saw H&S as the domain of contractors only. The findings of research conducted all over the world, particularly in Europe and the United Kingdom, are well documented and show that up to 67% of the causes of accidents that were investigated occurred upstream from the involvement of contractors and ranged from client choices and poorly designed features in the building. In the EU, for example, the Commission found that more than 50% of work-related accidents on construction sites were because of unsatisfactory architectural and/or organisational options, or poor planning of the works at the early project preparation stage. The Commission also found that a significant number of accidents stemmed from inadequate coordination, especially where various undertakings worked simultaneously or in succession at the same construction site. The accidents that occurred were accidents just waiting to happen as soon as the contractor began working on the construction site. Consequently, there has been a major paradigm shift to redistribute the responsibility for construction worker health and safety away from the contractor, who was previously solely responsible, to include all participants in the construction process, from the client through to the end-user. However, given that the contractors are involved in more aspects of the actual construction process than other stakeholders, the responsibility on contractors to contribute to a team approach to construction H&S is not insignificant.

Within contractor organisations leadership, commitment and involvement in construction H&S, particularly of senior or top management, are the most important elements at the organisation level, whereas risk assessment and management are most important at the project level. Further, contractors should cultivate their own charisma and their ability of being positively influential about the attitudes towards H&S they would want to see evidenced in their project and construction managers and require them to behave as role models for others in the organisation. Several international studies identified the following as being typically characteristic in contractors when it came to H&S:
- Lack of management commitment and involvement in H&S
- Poor H&S performance caused by extensive subcontracting
- Lack of proper supervision
- Absence of adequate H&S training resulting in differences in training and competence levels on construction sites
- Absence of dedicated and qualified H&S personnel on site
- Ineffective H&S laws and lack of enforcement

- Extensive use of foreign workers to overcome both skills shortages and high wage demands by local workers
- Lack of awareness and taking responsibility for H&S by workers
- Inadequate work procedures
- Poor accident record keeping
- Lack of management commitment to H&S budget allocation.

The early involvement of contractors in project decision-making, including any decision-making that occurs before the commencement of the construction phase on site, is linked to the adoption of 'higher order' H&S risk controls.

It makes good business sense to include contractors as early as possible in the construction process because of the high levels of construction expertise they possess, especially because of their specialised training, knowledge and years of experience in the application of construction materials, means and methods. It is likely that if involved in the early stages of a project, contractors will be able to provide advice during 'upstream' decision-making, particularly during the planning and design stages of a project, about various H&S hazards and the related risks of exposure and the ways to mitigate them in construction activities. Arguably, this amount of knowledge and information during the early stages of a construction project is likely to improve the effectiveness of H&S in design activities and facilitate the adoption of technological controls for H&S risks on the construction site. It is essential that contractors have a comprehensive and sound understanding of construction H&S requirements, knowledge of all applicable laws and regulations, and a commitment to continuous H&S improvement. Arguably, once contractors believe that all accidents are avoidable, huge improvements are possible.

Contractors frequently argue that making financial provision for H&S in their tenders or bids will make them uncompetitive and lose competitive advantage. This is a short-sighted view. In the longer term it is the contractor who conscientiously and deliberately develops a working environment that is visibly safe and without any threat to the health, not only of its workers, but its entire supply chain that will become the preferred contractor of choice. Although the price paid for H&S measures may seem high to some clients, in the longer term it should be off-set by lower tender prices as contractors experience savings associated with better construction H&S. Arguably, the primary areas of savings are reduced insurance premiums, less disruption of work schedules and programmes, and higher levels of worker productivity as workers feel more safe and secure on sites. Benefits accruing to the financiers of projects include lower credit risk, less likelihood of work stoppages and diminished risk to their reputation.

In South Africa clients in their tender or bid documents are required to allow the prospective contractor to make financial provision for the mitigation of all residual significant hazards, included from the baseline H&S risk assessment of the client that was done initially, and considered and incorporated in the design of the structure to be erected. Rather than the client just allowing a provisional sum or a uniform allowance eliminating the H&S competitive advantage of various competitors in the bidding process, provision should be made, for example, for items that could be separately priced, such as the following:

- The preparation and updating of a site-specific H&S plan, including a supervision and reporting scheme to include subcontractors or co-contractors
- Provision of temporary protective structures such as scaffolds and hoardings
- Hiring of a qualified H&S officer and other H&S personnel
- Provision of appropriate and relevant H&S training to workers and supervisors
- Cost of arranging and attending meetings of the H&S committee
- Provision of welfare facilities such as water, food, housing, ablutions with running water, hand sanitisers, masks and social distancing requirements
- Provision of personal protective equipment
- Medical examinations and first aid
- Emergency facilities.

In Hong Kong, for example, the cost of meeting the H&S requirements of the client was taken out of the competition by pre-pricing these items instead under the 'Pay for Safety' scheme. In terms of this scheme, the maximum payment for all H&S items was set at approximately 2% of the estimated value of the contract on small projects and 1% on large projects. Items that were not delivered by the contractor were not paid for when the final account of the project was computed.

4.6 PRINCIPAL CONTRACTORS

In terms of South African regulations, contractors must specifically do the following:
- Provide and demonstrate to the client a suitable, sufficiently documented and coherent site-specific H&S plan, based on the documented H&S specifications of the client, which plan must be applied from the date of commencement of and for the duration of the construction work and which must be reviewed and updated by the principal contractor as the work progresses.
- Open and keep on site an H&S file as part of the H&S management system and record keeping[5] that must include all documentation required in terms of the Act and regulations and must be made available on request to an inspector, the client, the client's agent or another contractor working on the project.

On appointing any other contractor, the principal contractor must do the following:
- Provide contractors who are tendering to perform construction work for the principal contractor with the relevant sections of the H&S specifications pertaining to the construction work that has to be performed.

---

5   Note that this file is to be compiled during the execution of the project and contain all relevant records for managing construction H&S effectively.

- Ensure that potential contractors submitting tenders have made sufficient provision for H&S measures during the construction process.[6]
- Ensure that no contractor is appointed to perform construction work unless the principal contractor is reasonably satisfied that the contractor that he or H&S intends to appoint, has the necessary competencies and resources to perform the construction work safely.[7]
- Ensure prior to work commencing on the site that every contractor is registered and in good standing with the compensation fund, or with a licensed compensation insurer as contemplated in the Compensation for Occupational Injuries and Diseases Act, 1993.
- Appoint each contractor in writing for the part of the project on the construction site.
- Take reasonable steps to ensure that each contractor's H&S plan is implemented and maintained on the construction site.
- Ensure that the periodic site audits and document verification are conducted at intervals mutually agreed upon between the principal contractor and any contractor, but at least once every 30 days.
- Stop any contractor from executing construction work that is not in accordance with the H&S specifications of the client and the H&S plan of the principal contractor for the site or that poses a threat to the H&S of persons.
- Where changes are brought about to the design and construction, make available sufficient H&S information and appropriate resources to the contractor to execute the work safely.[8]
- Discuss and negotiate with the contractor the contents of the H&S plan and thereafter finally approve that plan for implementation.
- Ensure that a copy of the H&S plan as well as the H&S plan of the contractor is available on request to an employee, an inspector, a contractor, the client or the client's agent.

---

6   Sufficient provision implies that in the same way as the principal contractor has to show the financial provision for the costs of H&S, the contractor and sub-contractors must do the same and the principal contractor must make the necessary provisions in bid documentation or work packages.

7   By implication a process of prequalification of the contractors to be considered must be done by the principal contractor.

8   Contractor must be allowed to price for any additional H&S measures as a result of these changes.

- Hand over a consolidated H&S file[9] to the client upon completion of the construction work, which includes a record of all drawings, designs, materials used and other similar information concerning the completed structure.
- In the H&S file[10] include and make available a comprehensive and updated list of all the contractors on site accountable to the principal contractor, the agreements between the parties and the type of work being done.
- Ensure that all his or her employees have a valid medical certificate of fitness specific to the construction work to be performed and issued by an occupational health practitioner.

4.7 CONTRACTORS

A contractor must do the following prior to performing any construction work:
- Provide and demonstrate to the principal contractor a suitable and sufficiently documented H&S plan, based on the relevant sections of the H&S specification of the client and provided by the principal contractor, which plan must be applied from the date of commencement of and for the duration of the construction work and which must be reviewed and updated by the contractor as the work progresses.
- Open and keep on site an H&S file,[11] which must include all documentation required in terms of the Act and these regulations, and which must be made available on request to an inspector, the client, the client's agent or the principal contractor.
- Before appointing another contractor to perform construction work be reasonably satisfied that the contractor that he or H&S intends to appoint has the necessary competencies and resources to perform the construction work safely.
- Cooperate with the principal contractor as far as is necessary.
- As far as is reasonably practicable, promptly provide the principal contractor with any information which might affect the H&S of any person at work carrying out construction work on the site, any person who might be affected by the work of such a person at work, or which might justify a review of the H&S plan.

---

9   There is confusion in South Africa about what this H&S file is to contain. The H&S file referred to earlier (and the use of the term 'file' is unfortunate) is for use DURING the construction phase of the project and must contain all relevant information for the purpose of managing H&S during this period. The H&S file to be handed over to the client at the end of the project is for use AFTER the construction phase (post-construction), for the purposes of maintaining the facility and providing the services in the facility and is useful should any changes need to be made in the form of alterations, especially structural ones. Therefore, this file or record should contain ONLY the information that will be useful post-construction as part of facilities and maintenance management of the completed facility – like the manual of a car when one purchases one.

10   This point highlights the need for this information post-construction so that should any defects manifest or additional work need to be done the information will assist in that process.

11   Note that this file is to be compiled during the execution of the project and contain all relevant records for managing construction H&S effectively.

4.8 CO-CONTRACTORS

To complete and deliver a construction project successfully demands of necessity a high level of teamwork with shared values, goals and objectives synchronous with those of the client. The use of the term 'sub-contractor' in the industry is an unfortunate one. The prefix *sub* is by definition indicative of 'something being less than'. Surely the intention is not that 'sub-contractors' are less than contractors. Unfortunately, the way they are treated by principal or main contractors, both in practice and in many standard forms of contract, suggests that the opposite is true. They frequently bear the brunt and burden of the non-performance of the principal contractor without the same contractual protection with respect, for example, to payment on time for the work they have completed. Arguably, it is time for the term 'sub-contractors' to be replaced by referring to them as 'co-contractors' because that is what they really are. They are part of the team that works together with the rest of the project team to complete and deliver construction projects successfully.

In South Africa, where a contractor appoints another contractor to perform construction work, the duties that apply to the principal contractor apply to the sub-contractor as if he or she were the principal contractor.

4.9 GENERAL REQUIREMENTS

Further to previous and other requirements, the following are applicable in South Africa:
- A principal contractor must take reasonable steps to ensure cooperation between all contractors appointed by the principal contractor to enable each of those contractors to comply with the construction regulations.
- No contractor may allow or permit any worker[12] or person to enter any site, unless that worker or person has undergone H&S induction training pertaining to the hazards prevalent on the site at the time of entry.
- A contractor must ensure that all visitors to a construction site undergo H&S induction pertaining to the hazards prevalent on the site[13] and must ensure that such visitors have the necessary personal protective equipment.
- A contractor must at all times keep up his or her construction site records[14] of the H&S induction training, and such records must be made available on request to an inspector, the client, the client's agent or the principal contractor.

---

12   Refers to construction workers on site, supervisors and all other employees of the construction organisation.
13   Both general pertaining to the project as well as the specific hazards that they might be exposed to on the particular day that they are on the site.
14   These can for practical purposes be kept in the H&S file used during the construction phase and excluded from the H&S file handed over to the client after construction has been completed.

- A contractor must ensure that all his or her employees have a valid medical certificate of fitness specific to the construction work to be performed and issued by an occupational health practitioner. This medical certificate of fitness must include the following information about the construction worker:
  - Employee name
  - ID number
  - Company number
  - Occupation
  - List of possible exposures
  - Job specific requirements
  - Any PPE needed
  - Certification by occupational health practitioner.

REVIEW QUESTIONS

1. What are the six construction project phases?
2. What contribution can clients make to construction health and safety?
3. What are the primary roles that designers should play on a construction project relative to construction health and safety?
4. What is the relationship between the principal contractor and other project stakeholders with respect to construction health and safety on their sites?
5. Describe the role and responsibilities of the client-appointed health and safety agent?
6. What is the responsibility of the SACPCMP regarding construction health and safety?

REFERENCES

Agumba, J & Haupt, T. 2014. A health and safety performance improvement model for small and medium enterprises in the South African construction industry. Proceedings of the 5th International Conference on sustainable built environment, Kandy, Sri Lanka, 12–15 December 2014

Agumba, J & Haupt, T. 2014. The types of accidents and injuries encountered by construction SMEs in South Africa. Proceedings of the 5th International Conference on sustainable built environment, Kandy, Sri Lanka, 12–15 December 2014

Agumba, J & Haupt, T. 2018. The influence of health and safety practices on health and safety performance outcomes in small and medium enterprise projects in the South African construction industry. *Journal of the South African Institute of Civil Engineering*, 60(3): 61–72

Akinlolu, M & Haupt, TC. 2020. Evolution in the intellectual structure of construction safety technology research: A bibliometric review. The Fifth Australasia and South-East Asia Structural Engineering and Construction Conference 30 November – 3 December, New Zealand

Akinlolu, M, Haupt, TC, Edwards, D & Simpeh, F. 2020. A bibliometric review of the current status, and emerging trends in construction safety management technologies. *International Journal of Construction Management*. doi.org/10.1080/15623599.2020.18 19584

Chigangacha, PS & Haupt, TC. 2017. Effectiveness of client involvement in construction projects: A contractor perspective. Proceedings of Association of Schools of Construction of Southern Africa (ASOCSA) 11th Built Environment Conference, 6–8 August, Durban, South Africa, pp 249–266

Deacon, C. 2020 Professional Construction Health and Safety Agent (CHSA) practice issues in the South African construction sector. In Aigbavboa C, Thwala W. (eds) *The Construction Industry in the Fourth Industrial Revolution*. CIDB 2019. Springer, Cham. Available at: https://doi.org/10.1007/978-3-030-26528-1_61

Haupt, TC. 2001. *The Performance Approach to Construction Worker Safety and Health.* Unpublished PhD dissertation, University of Florida

Haupt, TC. 2016. How do we compare? Presentation to Master Builders South Africa Congress, International Convention Centre, Durban, 31 August – 2 September 2016

Haupt, TC. 2017. Better health, safety, wellbeing, and life in construction. Presentation to Bahir Dar Institute of Technology, Ethiopia, on 25 August 2017

Haupt, TC. 2017. #Makeconstructiongreatagain# Better health, safety, wellbeing, and life in construction. Keynote address at 11th ASOCSA Built Environment Conference, Public Works Conference Center, Mayville on 7 August 2017

Haupt, TC. 2017. Pain and bane of H&S data. Presentation at Health and Safety Day, Master Builders Association KwaZulu-Natal, Durban, April 21, 2017

Haupt, TC. 2018. Pre-qualification criteria for Client Appointed Construction Health and Safety Agent. Presentation at Association of Construction Health and Safety Management Summit, Port Elizabeth, October 2018

Khoza, J & Haupt, TC. 2020. Measuring H&S performance of construction projects in South Africa. 14th Built Environment Conference (ASOCSA 2020), Durban, South Africa

Lim, SJ, Jeong, SC, Na, YJ & Won, JH. 2020. Analysis on construction clients' role for safety and health management in plan, design, and construction stage. *Journal of the Korean Society of Safety*, 35(3): 24–31

Lingard, H, Pirzadeh, P, Blismas, N, Wakefield, R & Kleiner, B. 2014. Exploring the link between early constructor involvement in project decision-making and the efficacy of health and safety risk control. *Construction Management and Economics*, 32(9): 918–931

Lingard, H, Wakefield, R & Walker, D. 2020. The client's role in promoting work health and safety in construction projects: Balancing contracts and relationships to effect change. *Construction Management and Economics*, DOI 10.1080/01446193.2020.1778758

Mahmoudi, S, Ghasemi, F, Mohammadfam, I & Soleimani, E. 2014. Framework for continuous assessment and improvement of occupational health and safety issues in construction companies. *Safety and Health at Work*, 5(3): 125–130

Musonda, I & Haupt, T. 2009. Construction public client health and safety culture in Botswana: A pilot study. In Haupt, T.C. (ed) The Built Environment 4, Proceedings of the 4th Built Environment Conference, Livingston, Zambia, 17-19 May, pp 283–292

Musonda, I & Haupt, TC. 2011. A Delphi study on the impact significance of clients and designers on health and safety (H&S) consideration throughout project lifecycle. *Journal of Construction*, 4(2): 3–7

Musonda, I & Haupt, T. 2011. An exploratory study on the impact significance of project stakeholders on construction project health and safety (H&S). 2011 International Conference on Construction Engineering and Project Management (ICCEPM), Sydney, Australia, 16-18 February

Musonda, I & Haupt, T. 2011. Conceptual model of client centered health and safety performance improvement. ASOCSA 6th Built Environment Conference, Sandton, South Africa, 31 July – 2 August, pp 156–167

Musonda, I & Haupt, T. 2011. Identifying factors of health and safety (H&S) culture for the construction industry. ASOCSA 6th Built Environment Conference, Sandton, South Africa, July 31-August 2, pp 112–127

Musonda, I, Haupt, TC & Smallwood, J. 2009. Client attitude to health and safety – A report on contractor's perceptions. *Acta Structilia*, 16(2): 69–87

Musonda, I, Pretorius, JH & Haupt, TC. 2012. Assuring health and safety (H&S) performance on construction projects: Clients' role and influence. *Acta Structilia*, 19(1): 71–105

Musonda, I, Thwala Didibhuku, W & Haupt, T. 2011. Impact significance of construction clients' culture on contractors' health and safety (H&S) culture – An exploratory Delphi Study. 6th International Conference on Construction in the 21st Century (CITC-VI), 'Construction Challenges in the New Decade,' Kuala Lumpur, Malaysia, 5–7 July

Na, YJ, Won, JH, Lee, SW & Yong Park, K. 2019. New trial for imposing mandatory role and responsibility of construction clients on construction safety and health management system in Korea

Okorie, V, Emuze, F, Smallwood, J & Van Wyk, K. 2014. The influence of clients' leadership in relation to construction health and safety in South Africa. *Acta Structilia*, 21(2): 44–68

Republic of South Africa. 2014. Construction Regulations. Government Printer.

Republic of South Africa. 2015. Notice regarding Application for Construction Work Permit of Construction Work in terms of the Construction Regulations 2014. Government Printer, Pretoria

SACPCMP. 2013. Registration Rules for Construction Health and Safety Agent (PrCHSA)

Smallwood, J & Deacon, C. 2017. The performance of construction health and safety agents. ISEC 2017 - 9th International Structural Engineering and Construction Conference: Resilient Structures and Sustainable Construction

Smallwood, J. 2019. Clients and construction health and safety (H&S). Proceedings of the Creative Construction Conference, 29 June - 2 July, Budapest, Hungary. Available at: https://doi.org/10.3311/CCC2019-090

Teo Ai Lin, E, Haupt, T & Feng, Y. 2008. Construction health and safety performance in developing and developed countries: A parallel study in South Africa and Singapore. Evolution of and Directions in Construction Safety and Health: Proceedings of CIB W99 14th Rinker International Conference, 9-11 March, 2008, Gainesville, Florida (pp 485-499)

Tymvios, N & Gambatese, JA. 2016. Perceptions about design for construction worker safety: Viewpoints from contractors, designers, and university facility owners. *Journal of Construction Engineering and Management*, 142(2): 04015078

Wells, J & Hawkins, J. 2014. Promoting construction health and safety through procurement: A briefing note for developing countries. Engineers Against Poverty, London

Wu, C, Wang, F, Zou, PX & Fang, D. 2016. How safety leadership works among owners, contractors and subcontractors in construction projects. *International Journal of Project Management*, 34(5): 789–805

CHAPTER 5
CULTURE AND LEADERSHIP

5.1 INTRODUCTION

South Africa is a rich multicultural and multilingual country with no fewer than 11 official languages. Some argue that if sign language is included then there would be 12 official languages. It would be foolhardy to ignore the impact of these cultural and linguistic differences on construction health and safety, especially when one considers the widely recognised transient nature of employment in the industry. Rarely are construction workers employed on a permanent basis, considering the changes in the structure of the industry and construction organisations in recent times. Most construction organisations outsource major portions of construction by means of sub-contracting arrangements which could be on a labour-only or labour-and-materials basis. Construction is universally accepted as a major provider of employment of labour and its strategic position and reputation as a sub-Saharan economic superpower makes South Africa an attractive destination for workers from other countries in the region to seek employment opportunities in the sector. These foreign workers bring with them their own culture and languages. Unless fully appreciated and understood, these cultural and language differences will be challenging to manage in the context of construction projects. In recent times, xenophobia has raised its ugly head creating tensions between foreign and local workers resulting in violence and criminal activity. These situations need to be managed effectively on construction projects.

5.2 CULTURE

While culture is a term that is frequently used, it is an extremely difficult concept to define as it has been deeply contested for many years. In South Africa culture in terms of what it is and what it means in practice is hotly debated. During the period between 1920 and 1950 there have been more than 150 attempts to define what culture is.

For practical purposes in the context of construction H&S, culture is defined as those individual patterns of behaviour, values and beliefs that impact how individuals think, decide and behave.

It is therefore likely that on any construction site there may be differing values among the labour force about work ethic, honesty, responsibility, race, gender and company loyalty – all of which need to be managed effectively to create a safe and healthy working environment. Other descriptions of what culture is include:

- the thinking of individuals that is produced by the words, ideas, symbols, noises and images with which they are confronted every day
- the meaning placed upon their experiences as individuals
- a consistent pattern of thought and action.

Therefore, the thought processes of construction workers who spend most of their time on construction sites are influenced by what they experience and see at work every day. If they see unchecked and ignored unsafe and unhealthy behaviour or conduct on the construction site, they are likely to behave and act in the same way and presume that such behaviour and actions are acceptable to the management of the organisation.

5.3 HEALTH AND SAFETY CULTURE

The concept of health and safety culture had its roots in several disciplines including psychology, sociology, anthropology and management. H&S culture in the construction context can be explained as being the product of the values, attitudes, beliefs, perceptions, understanding, values, competencies and patterns of behaviour of individual construction workers and the crews they work in that determine their commitment to the H&S management of the organisation. The prevailing H&S culture is reflected in the dynamic reciprocal relationships between the perceptions of the entire workforce about, and attitudes towards, the H&S goals of the organisation; the daily goal-directed behaviour and actions of the workforce; and the presence and quality of the H&S management system, audits and inspections. Therefore, H&S culture equates to the value that the organisation places on the H&S of its workforce through its policies, procedures and practices.

It has been found that construction organisations with a positive H&S culture are typically characterised by communications that are based on mutual trust, by shared perceptions of the importance of H&S and by confidence in the efficacy of the various preventive measures adopted by these organisations. As a minimum, H&S culture is reflected in the willingness of an organisation to at least develop and learn from errors, incidents and accidents if they prefer to be reactive rather being proactive.

On a practical level, construction H&S culture may be operationalised as:

- management commitment and involvement as it is defined at the group level or higher, which refers to the shared values among all the group or organisation members
- communication and feedback in multiple languages, forms of media and forums
- supervisory environment as it is concerned with formal H&S issues in an organisation and closely related to, but not restricted to, the management and supervisory systems

- supportive environment from all levels of management
- work pressures and stress determined by the preoccupation of the organisation with time, cost and quality
- personal appreciation of risk, because risk takers and risk avoiders see risk differently
- training and competence levels of the workforce
- H&S rules and procedures which have an impact on the behaviour of the workforce
- involvement and engagement of workers since it emphasises the contribution from everyone at every level of an organisation
- appraisal of workplace hazards and exposures in the form of regular HIRAs.

5.4 CONSTRUCTION HEALTH AND SAFETY CLIMATE

The concepts of construction H&S culture and climate have often been used interchangeably when they are, in reality, related but distinctly different. H&S climate is a leading indicator that can reveal gaps between management and construction worker perceptions, or between espoused and enacted policies, and trigger communication and action to narrow those gaps. It is an indicator of the overall nature and strength of the H&S culture of the organisation. H&S climate measurements can be used to proactively assess the effectiveness of a construction organisation in identifying and remediating work-related hazards, thereby reducing or preventing work-related ill health, injury and fatalities. It provides a means to understand and improve H&S on construction projects and construction sites because it gives insight into how healthy and safe construction workers perceive their work environment to be. It has been described as giving a 'snapshot' view of the H&S culture, policies, procedures and practices in an organisation.

Therefore, assessing H&S climate in an organisation can provide a reliable prediction of the level of the overall H&S culture of the organisation. It is predictive of workplace accidents, injuries, underreporting of H&S incidents, near misses, H&S knowledge, H&S motivation, H&S compliance and H&S-related organisational citizenship behaviours. H&S climate measurements can be used to proactively assess the effectiveness of an organisation in identifying and remediating work-related hazards, thereby reducing or preventing work-related ill health and injury. Therefore, it can be argued that the H&S climate in an organisation can benefit industry stakeholders by providing them with the knowledge of attitudes and perceptions that can help to consistently achieve better H&S performance. It can positively influence the H&S knowledge, motivation, attitudes and behaviours of workers, as well as reduce injury outcomes especially considering that H&S climate reflects the shared perceptions among the workforce regarding what is rewarded, expected, valued, and reinforced in the workplace with respect to H&S. It mirrors their understanding about the comparative significance of H&S performance in their job-related attitudes and behaviour.

Because the H&S climate reflects the extent to which the workforce perceive that H&S is prioritised within the workplace it allows them to ascertain the actual and real relevance of H&S within the organisation rather than what is purported to be relevant. The reality might be markedly different from the intention. A positive H&S climate provides the workforce with tangible and visible evidence that H&S behaviours and outcomes are valued, supported and rewarded in the workplace. On the other hand, a job site that is in continual flux, a high degree of sub-contracting, a transient workforce and individual work ethics impact negatively on the H&S climate in the organisation.

The most common indicators that are typically used to measure H&S climate include H&S policies, procedures and practices followed by general management's visible and tangible commitment to and involvement in H&S. The construction industry has largely relied on lagging indicators such as various ratios and indices to show the state and health of their overall H&S performance. However, these indicators are reactive and do not prevent negative consequences of poor H&S performance before they occur. Several studies have been done in different parts of the world to identify the dimensions to be included in H&S climate tools. The common H&S climate dimensions, which are a mix of lagging and leading indicators used across studies, are described in Table 5.1.

Table 5.1: Common health and safety climate dimensions

Dimension	Description
Management commitment to H&S	Refers to how effective top management members are in ensuring that H&S is a priority in their organisation
Supervisory H&S response	Refers to how responsible first-line leaders are regarding the implementation of organisational H&S procedures during day-to-day activities
H&S rules and procedures	Refers to the degree to which workers believe and follow the H&S rules and procedures of their organisation to prevent accidents/incidents
Communication	Refers to how members of both top management and front-line workers communicate H&S issues, and how openly managers receive feedback from workers about their H&S concerns
Worker involvement	Refers to the degree to which workers receive encouragement from the upper management to participate in H&S procedures and the extent to which they are invited to be a part of policy creation
Training	Refers to the amount of H&S education and instruction that workers receive during their work
Risk-taking behaviour	Refers to the degree of risk that workers are willing to take to complete tasks while violating H&S regulations in the organisation
Workload pressure	Refers to the amount of work that leads workers to perform work in unsafe and unhealthy ways

These studies found that management commitment to H&S, supervisory H&S rules, H&S rules and procedures, training, and individual responsibility of H&S were significantly correlated with worksite injuries. Where these were poor, more worksite injuries occurred.

However, the focus instead should be on leading indicators which precede consequences and can drive activities and interventions that prevent and control the occurrence of any negative consequences and foster a positive H&S climate. With this approach in mind, the Center for Construction Research and Training (formerly the Center to Protect Workers Rights (CPWR)) in the USA has developed a model Safety Climate Assessment Tool (S-CAT) that can be freely used by the industry at no charge to improve leading indicators and thereby overall jobsite H&S climate. This model has been adapted with permission from CPWR to suit the South African construction H&S environment in the form of the South African Safety and Health Climate Assessment Tool (SASH-CAT).

Whereas the CPWR model comprises eight leading indicators, the SASH-CAT comprises the following nine leading indicators:
1. Demonstrating management commitment and involvement (6 items)
2. Aligning and integrating H&S as a value into the business (6 items)
3. Ensuring responsibility and accountability at all levels (4 items)
4. Improving supervisor leadership (3 items)
5. Empowering, involving and engaging workers (3 items)
6. Improving communication (3 items)
7. Training at all levels (6 items)
8. Encouraging client involvement (6 items)
9. Responsive emergency and disaster management (x items)

A rubric indicative of five possible levels of maturity of the organisation relative to each indicator is used. The maturity levels are as follows:
1. Ad hoc
2. Reactive
3. Compliant
4. Proactive
5. Better practice

These maturity levels together with examples of descriptors are shown in Table 5.2.

Table 5.2: Levels of maturity of an organisation and examples of descriptors

Ad hoc	Reactive	Compliant	Proactive	Better practice
Management representatives rarely visit the actual job site. When they do visit they often act as poor H&S role models by breaking the H&S policies and procedures of the organisation. Management does not participate in H&S audits. When workers raise any H&S issues and concerns to any level of management they are ignored and not acted upon.	Management gets involved only after an injury occurs. They often blame workers for injuries, leading to censures that may include termination. H&S rules are enforced only after an incident, accident or a fatality or when audit findings are negative.	Management conforms strictly to the OHS Act, Construction Regulations and other legislation and regulations. H&S compliance is based on client or regulatory directives and requirements. Managers participate in H&S audits.	Management initiates and actively participates in H&S audits. Managers regularly meet with workers to ask for advice and feedback regarding hazard reduction. Management conducts spontaneous site visits and recognises workers for identifying hazards, working safely and keeping co-workers safe. Managers participate in H&S programme development and provide adequate resources to ensure a positive H&S climate. The H&S management system is reviewed annually to ensure effectiveness and relevance.	Management integrates H&S into every meeting and engages in continuous improvement regarding H&S conditions and hazard reduction. External audits are conducted to evaluate the involvement of senior management in H&S. Managers are held accountable for H&S expectations through annual performance evaluations. H&S trends are analysed. There is a formalised review process for monitoring and evaluating corrective actions.

The extent of the commitment of the organisation to adopt or implement each item that makes up a particular leading indicator, together with an indication of the time period within which it is intended to do so, can be rated using the following five options:

1. **Current** implementation in the organisation
2. **Short-term** implementation, say within the next 1 or 2 months
3. **Medium-term** implementation, say between 6 and 12 months
4. **Long-term** implementation, say between 1 and 2 years
5. **No** implementation

The rating for each leading indicator is calculated by taking the mean of their respective items, for example, the six items of the indicator for demonstrating management commitment and involvement. A SASH-CAT rating for the organisation is computed by calculating the total mean of all the items. This rating is indicative of the H&S climate within the organisation and provides an indication of areas that can be improved. This rating can be plotted graphically using a spider diagram.

From the SASH-CAT tool a detailed report can be generated that provides detailed feedback on how the organisation scores on each H&S climate factor. Once the tool has been used in a number of organisations it will be possible to benchmark the H&S climate in the organisation against other competitors in the industry. With the ratings of the SASH-CAT as a basis, organisations can be provided with recommendations and suggestions for how to improve in each area that is deficient. It is therefore possible for an organisation to monitor and evaluate the effectiveness of any H&S interventions introduced into the organisation based on its SASH-CAT feedback.

5.5   LANGUAGE AND COMMUNICATION

The effect of language on H&S culture cannot be ignored, especially in South Africa with its many languages, all of which could be spoken on the same construction site. Communication is a two-way process that involves the sending and receiving of symbols, signs or signals in the form of, for example, words, pictures, things and actions. Communication has been defined as simply: 'what we do to give and get understanding'. Communication has also been described as an action that provides information that is relevant and meaningful to persons receiving the information. This information might not have the same meaning to different people and may not produce the same outcomes that are typically manifested in behaviour and actions. Linguistic relativity in terms of the Sapir-Whorf hypothesis suggests that speakers of different languages think differently because of the differences in the construction and functioning of their languages. Therefore, language difficulties create a communication barrier that can lead to confusion and misunderstanding with negative outcomes for construction H&S. Typically, construction workers are confronted with language challenges which include the use of colloquialisms, low levels of education, and cultural work environments where minimal emphasis is placed on H&S.

The accuracy and consistency of the words and symbols used to communicate H&S information is important to achieve the desired mutual understanding. The success and effectiveness of H&S communication involves the correct use of language that includes words, pictures and body language, and how these are received and interpreted. It is speaking and listening, clarifying and explaining, writing and reading, behaving and observing behaviour, reinforcement and feedback. Its goal is to achieve understanding. In a multilingual country like South Africa with its 11 official languages besides colloquial slang, achieving this goal is more challenging. These differences in language must be accommodated in the forms of H&S-related communication. Language used in the workplace or on the factory floor is different from that used in an office environment as workers create their own work-related vernacular or colloquial slang to communicate among themselves, and that vernacular should be recognised.

Furthermore, information that is shared must account for the capability of the worker to process the information itself as well as the volume of it. A phased approach over time is advocated using different communication methods that include reiteration and repetition to avoid boredom, monotony, and apathy. Correct interpretation of information received is dependent on the personal experiences, previous knowledge, schooling, training, attitudes and emotions of the worker.

Figure 5.1: Using signs to communicate construction health and safety information regardless of the language spoken

Increasingly today social media platforms such as WhatsApp chat groups, Facebook groups, Instagram and similar evolving platforms and apps are proving to be popular and effective forms of communication and should be considered as part of H&S communication strategies.

Opportunities to communicate H&S include, for example, training practices that include orientation or induction and regularly scheduled training, toolbox talks, informal communication in the form of, for example, pamphlets and formal presentations on H&S. It is important to consider that implementing an appropriate language and literacy H&S training programme can prevent incidents, accidents, injuries and fatalities in the workplace. Construction workers who receive inadequate H&S training or have limited understanding of potential hazards on the job site are at risk of work-related incidents, accidents, injuries or fatalities. Therefore, describing hazards to workers whose first language or mother tongue is not English, can be done making use of pictograms, audio-visual material and practical demonstrations. Other strategies such as hands-on H&S training and colour-coded signs can communicate H&S training information to construction workers with language barriers, low literacy skills and cultural differences.

5.5.1 PERSONAL COMMUNICATION

Personal communication normally occurs in the following five key areas of day-to-day responsibilities at work:
1.   Individual job or work activity and workplace orientation or induction
2.   Job/task/work activity instruction
3.   Planned personal contacts
4.   Key point tips whenever needed
5.   Job performance coaching, which is a more recent type of targeted personal communication.

INDIVIDUAL JOB ORIENTATION

Numerous studies show that new construction workers are almost twice as likely to have an accident on a construction site as experienced workers. This phenomenon also applies to workers who are being transferred between various construction sites. Organisations with formal orientation or induction for all new workers and transferees have on average workers' compensation 'modification rates' of around 25% lower than organisations that do not have any formal orientation or induction programmes for new workers and transferees.

Workers who are new to the job and to the work environment are especially at risk. Good and effective supervisors use proper orientation or induction to help new or transferred workers get safely through that critical period without harm.

Good orientation or induction requires empathy, the ability to be in the shoes of the workers, and to see things from their viewpoint. It should not merely be a one-sided presentation of facts and figures about the organisation. Rather it should be a two-way process of reaching mutual understanding, of welcoming the new workers or transferees into the organisation or on the particular job site, helping them to become familiar with the work environment and laying the basis for the desired H&S knowledge, skills and attitudes.

Every supervisor must be conscious that first impressions tend to be lasting impressions, and that no one ever gets a second chance to make a good first impression. The orientation or induction process provides a unique opportunity to start building a solid supervisor–worker relationship, to let the behaviour of supervisors show their care and their concern for the new workers or transferees. It allows supervisors to determine what kind of mental image they want workers to have of them and then work to establish that very image.

TASK INSTRUCTION

Proper job or task or work activity instruction involves getting workers to execute assigned construction tasks correctly, quickly, conscientiously and safely. Effective instruction is a systematic and effective substitute for learning by trial-and-error and a reliable replacement for instruction that is hit-or-miss or random. The two basic goals of task instruction are:

1. To help motivate the worker to do the assigned task or work activity properly and safely.
2. To ensure that the worker knows how to do the job properly and safely.

KEY POINT TIPS

'Key points' are pieces of information that could make or break the job. They are special 'tricks of the trade' that could make the construction task more efficient. They are critical points of quality, productivity, cost control or health and safety.

'Tip' refers to a piece of information given in an attempt to be helpful; a small gift; a hint or suggestion. Key point tipping, therefore, is the organised process of giving construction workers helpful hints, suggestions, reminders or tips about key quality, production, cost or health and safety points in their work.

The best tips are short tips which should always be given as reminders rather than formal instruction. The following are some examples:

- Health and safety tip — 'Skosana, will you ensure that you always keep the guard of that circular saw in place as it would be tragic if you lost a finger if you did not?'
- Production tip — 'Koos, you can increase the rate at which you lay bricks if you had stacks of bricks laid out along the length of wall you are building.'
- Cost tip — 'Piet, will you go around the site every day and pick up all the nails that are lying around? Not only will the site be safer, but those nails can be reused and save the business some money.'
- Quality tip — 'Siya, you know that you always have time to do something over but never enough time to do it right the first time. Let's change that today!'

JOB PERFORMANCE COACHING

Job performance coaching is the day-to-day actions taken to help construction workers perform as best as they can. Effective coaching is 'leadership in action' and involves motivating, communicating and developing workers. It is 'management control in action', built on the activities of work identification, standards, measurement, evaluation, correction and commendation. Effective job coaching is an excellent means for building better relationships between supervisors or managers and their construction workforce.

Each construction worker must know what is expected of them, how they are doing and what they should do to improve. Typically, workers look to the supervisor or manager for guidance. Supervisors or managers must provide that guidance and lead construction workers to a winning performance on an efficient, safe, healthy and productive job site. Job performance coaching is based on the simple, basic principle that every worker has a right to know:

- what their job is
- the job performance metrics
- how they are doing
- specific steps to improve their performance.

When construction workers do not know these, the consequences are almost always unpleasant. For example:

- When workers do not really know what their jobs are, the results are confusion, wheel spinning, and losses such as waste, damage and injuries.
- When workers do not know their performance metrics or goals, the results are guesswork, wrong priorities, misdirected energies or hit-and-miss efforts.
- When workers do not know how they are doing, the results are low motivation, poor morale, and measurement by mind reading.
- When workers do not know specifically how to improve, the results are glittering generalities, procrastination and maintaining the status quo.

5.5.2 EFFECTIVE COMMUNICATION

The importance of effective communication on the job site cannot be overemphasised. It is important in problem solving, conflict resolution, for positive working and personal relationships, and in reducing the stresses associated with interpersonal interactions. Table 5.3 shows some examples of behaviours and phrases that often stop a person from communicating and which should be avoided if effective communication is to happen.

Table 5.3: Examples of behaviours and phrases that hamper communication

Ordering	'Don't talk like that.'
Warning	'If you do that, you'll be sorry.'
Moralising	'You ought/should ...'
Advising	'I suggest that you ...'
Reasoning with	'Let's look at the facts.'
Diagnosing	'You feel that way because ...'
Judging	'You are wrong about that.'
Name calling	'You are acting like ...'
Distracting	'Let's talk about something else.'
Interrupting	'But what about ...'

5.6 LEADERSHIP

The impact of management or organisation leadership on overall H&S performance is well established. Where top or senior management report a strong commitment to and involvement in construction H&S it is associated with higher overall H&S performance. However, to deliver complex construction projects such as major mega-infrastructural installations in remote areas using designs and technologies not previously encountered without threats to the health and safety of all stakeholders requires leadership and not just good management. It is important to note that leadership is NOT management. Arguably, the main difference between leaders and managers is that leaders have people follow them while managers have people who work for them. Leadership is having a vision, sharing that vision and inspiring others to support that vision while creating their own vision. It has also been described as the art of serving workers by equipping them with training, tools and human resources as well as the time, energy and emotional intelligence so that they can realise their full potential in the workplace. Nelson Mandela suggested that a person must follow the dictates of their conscience irrespective of the consequences which might overtake them for it. Leadership, therefore, is important in the creation of a culture that supports and promotes a strong H&S performance in an organisation.

Some differences between leaders and managers, adapted from the widely cited management theory of Warren Gamaliel Bennis, an American scholar, are shown in Table 5.4.

Table 5.4: Some differences between leaders and managers

The manager	The leader
• Administers	• Innovates
• Reacts to change	• Creates change
• Maintains	• Develops
• Focuses on systems and structure	• Focuses on people
• Relies on control	• Inspires trust
• Has short-range view	• Has long-range perspective
• Communicates	• Persuades
• Asks how and when	• Asks what and why
• Eye always on the bottom line	• Eye is on the horizon
• Has good ideas	• Implements ideas
• Accepts the status quo	• Challenges the status quo
• Is the classic good soldier	• Is his or her own person
• Directs groups	• Creates teams
• Takes credit	• Takes responsibility
• Is focused	• Creates shared focus
• Exercises power over people	• Develops power with people

According to Bennis, true leaders understand themselves, possess both a vision and the ability to translate that vision to their teams and are able to establish an environment of trust.

Ethical leadership is the demonstration of normatively appropriate conduct through personal actions and interpersonal relationships and the promotion of such conduct to followers through two-way communication, reinforcement, and decision-making. It is generally accepted that there are two types of leadership styles, namely transformational and transactional leadership. Transactional leadership emphasises effective and efficient task organisation and reliable task accomplishment. On the other hand, transformational leadership focuses on inspiring workers to go above and beyond their own self-interests for the good of the entire organisation. It is essential in construction H&S management as it is associated with positive H&S outcomes, such as an improved H&S climate, increased H&S behaviours and attitudes, and decreased numbers of accidents and injuries Transformational leaders foster closer relationships with their workers. Typically, transformational supervisors inspire their construction workers using the following tactics:

- Serving as a role model for the desired H&S behaviour
- Developing the H&S goal commitment of their workers
- Intellectually engaging workers in problem solving
- Empathising with their workers through individualised consideration
- Practising management by exception, which gives workers the responsibility to make decisions and fulfil their work or projects by themselves

- Allowing construction workers the freedom to do what they want as long as they get the work done correctly
- Rewarding workers when the set H&S goals are accomplished on time, or ahead of time.

Some of the benefits of H&S leadership include the following:
- Reduced number of lost time accidents
- Improved reporting of minor accidents, incidents and near misses
- Timeliness of delivery and quality of product
- Improved resource management
- Innovative solutions
- Enhanced visibility and transparency of purposes
- Improved overall project performance
- Opportunity to capture lessons learned
- Integrated project team
- Improved risk management
- Identifiable project goals and successes.

5.6.1 PRINCIPLES OF HEALTH AND SAFETY LEADERSHIP

Some of the principles of effective H&S leadership include the following:
- Commitment *and* involvement
- Being aware of the risks associated with construction activities
- Understanding and validating how these activities are supervised and executed
- Learning from incidents and responding appropriately
- Understanding and implementing appropriate H&S indicators
- Regularly reviewing H&S performance, document findings and making appropriate adjustments
- Being relentless.

According to the New South Wales Mine Safety Advisory Council in Australia, there are a number of 'Platinum' rules of H&S leadership, which include the following:
- Being aware of working with people
- Listening aggressively and talking to construction workers
- Walking the talk
- Taking command
- Addressing concerns promptly
- Avoiding unnecessary paperwork
- Improving the H&S competency of everyone
- Encouraging workers to give bad news
- Creating a climate of trust
- Moving from 'I' to 'We'
- Fixing own workplaces first – lead by example
- Proactively measuring and monitoring risks

- Regularly reviewing H&S management systems
- Allocating adequate resources in terms of time and money.

5.6.2 CHALLENGES OF HEALTH AND SAFETY LEADERSHIP

On a practical level there could be several challenges that impact effective H&S leadership. These include, for example:
- Not perceiving there to be a need
- Concerns about what might be discovered
- Resource issues
- Difficulties with making long-term commitments
- Belief that there is nothing positive to achieve
- Outcomes of reviews might be worse than those of competitors
- Unclear lines of communication and responsibility
- Organisational bureaucracy.

REVIEW QUESTIONS

1. Why is effective communication so important?
2. How would you improve your communication with workers in your workplace?
3. Why is it important to be an active listener?
4. What would be some of the pitfalls to avoid that negatively affect communication at work?
5. How would you use various ICT platforms to communicate within your organisation in the event of a national disaster such as the recent COVID-19 pandemic?

REFERENCES

Alruqi, WM, Hallowell, M & Techera, U. 2018. Safety climate dimensions and their relationship to construction safety performance: A meta-analytic review. *Safety Science,* 109: 165–173

Anderson, DS & Miller, RE. 2017. *Health and Safety Communication: A Practical Guide Forward.* Routledge (Taylor and Francis Group)

De Jesus-Rivas, M, Conlon, HA & Burns, C. 2016. The impact of language and culture diversity in occupational safety. *Workplace Health & Safety,* 64(1): 24–27

Feng, Y, Teo, EAL, Ling, FYY & Low, SP. 2014. Exploring the interactive effects of safety investments, safety culture and project hazard on safety performance: An empirical analysis. *International Journal of Project Management,* 32(6): 932–943

Gunduz, Z. 2017. A critical approach to culture and society definitions. *PEOPLE: International Journal of Social Sciences,* 3(2): 946–964

Haupt, TC. 2016. Leadership in H&S. Presentation at Association of Construction Health and Safety Management Summit, Westville on 10 October 2016

Haupt, TC. 2021. *Management of Safety, Health and Environment in South Africa: A Handbook.* Newcastle upon Tyne: Cambridge Scholar Publishing

Haupt, TC, Munshi, M & Smallwood, J. 2004a. Combating HIV and AIDS in South African construction through effective communication. *Acta Structilia*, 11(1 & 2): 26–43

Haupt, TC, Munshi, M & Smallwood, J. 2004b. The role of communication in combating HIV and AIDS in construction. *Revista Ingenieria De Construccion*, 19(2): 111–120

Lin, S, Mufidah, L & Persada, SF. 2017. Safety culture exploration in Taiwan's metal industries: Identifying the workers' background influence on safety climate. *Sustainability* 9(11): 1965. Available at: https://doi.org/10.3390/su9111965

Lowe, KB, Kroeck, KG & Sivasubramaniam, N. 1996. Effectiveness correlates of transformational and transactional leadership: A meta-analytic review of the MLQ literature. *Leadersh. Q*, 7: 385–425

NIFC. nd. *Keys to effective communication.* Available at: https://www.nifc.gov/hrsp/tools/keys_to_effective_communication.pdf

Nkolimwa, D, Jani, D & Dominic, T. 2020. Management practices on occupational health and safety in the Tanzanian's small scale mining firms: Does compliance cost matter? *Business Management Review*, 22(2)

Probst, TM, Goldenhar, LM, Byrd, JL & Betit, E. 2019. The safety climate assessment tool S-CAT: A rubric-based approach to measuring construction safety climate. *Journal of Safety Research*, 69: 43–51

Safeopedia. *What is the difference between a safety culture and a safety climate and why does it matter.* Available at: https://www.safeopedia.com/what-is-the-difference-between-a-safety-culture-and-a-safety-climate-and-why-does-it-matter/2/8059

Schwatka, NV, Hecker, S & Goldenhar, LM. 2016. Defining and measuring safety climate: A review of the construction industry literature. *Annals of Occupational Hygiene*, 60(5): 537–550

Shen, Y, Ju, C, Koh, TY, Rowlinson, S & Bridge, AJ. 2017. The impact of transformational leadership on safety climate and individual safety behavior on construction sites. *International Journal of Environmental Research and Public Health*, 14(1): 45

Teo, EAL & Feng, Y. 2009. The role of safety climate in predicting safety culture on construction sites. *Architectural Science Review*, 52(1): 5–16

University of Wisconsin-Milwaukee nd. *Principles of Communication.* [online] www4.uwm.edu. Available at: https://www4.uwm.edu/cuts/bench/commun.htm#princ

Vigorosa, L, Caffaro, F & Cavallo, E. 2020. Occupational safety and visual communication: User-centred design of safety training material for migrant farmworkers in Italy. *Safety Science*, 121: 562–572

Wu, C, Wang, F, Zou, PX & Fang, D. 2016. How safety leadership works among owners, contractors and subcontractors in construction projects. *International Journal of Project Management*, 34(5): 789–805

CHAPTER 6
MANAGING CONSTRUCTION
HEALTH AND SAFETY

6.1   INTRODUCTION

Around the world the construction industry is worth an estimated USD12.7 trillion, accounting for about 14% of the world's economy, and therefore is a significant industrial sector. Many reasons have been given for the poor state of construction H&S and the unacceptably high incidence of accidents, injuries and fatalities that occur frequently on construction sites. According to the International Labour Organization, each year there are about 60,000 fatal accidents on construction sites around the world. This is one fatal accident every ten minutes. It is the fourth most dangerous industry with the second most fatal injuries of all industrial sectors. The reasons for this poor performance include, for example, the labour-intensive nature of construction activities, the dynamic nature of construction, the uniqueness of the products of construction in terms of their shape or aesthetic appearance, their size and purpose, inadequate construction H&S legislative and regulatory framework, separation of design from construction in the construction process, poor or lack of construction H&S education and training, and lack of visible commitment to and tangible involvement of senior or top management in H&S on their construction projects and sites. Arguably, it is still possible, despite these impacts, to achieve better overall quality construction H&S performance in the industry. Therefore, for construction H&S to be effectively managed it needs to be integrated into the overall management of the construction project and involve all project stakeholders to work together through all the construction project phases to achieve the goal of a zero incident, injury, accident and fatality project and industry.

The perpetual poor H&S performance of the construction sector places a heavy burden on society at large, especially the large number of fatalities and physical disablements. Strong construction H&S management will help construction organisations reduce accidents and ill health, avoid costly prosecutions, reduce insurance costs as well as creating a culture of positivity in the organisation.

However, before discussing how construction H&S management can be improved, it is necessary to understand what construction safety is and what construction health is. They are distinctly different despite some arguing that construction health is a subset of construction safety and therefore one only needs to refer to construction safety. If that were true, there would not be any need for the Occupational Health and Safety Act. Instead, an Occupational Safety Act would suffice.

6.2 CONSTRUCTION SAFETY

Construction site safety is an aspect of construction that is concerned with protecting construction site workers and others such as the project team members and the general public from death, injury, disease or other health-related risks that may arise from construction activities. Construction safety generally refers to the condition of being protected, or safe from hazards, perils and other undesirable events during the construction process. Safety can also be used to refer to how safe or protected against harmful events a building or structure is when it is under construction or in use such as extreme weather conditions, security, operational failures or hazards. According to The British Standards Institution, in 2021, **building or construction safety** is defined as a '… matter relevant to protecting the safety of people from risk in and around buildings including but not limited to fire safety, structural safety, public safety and pertaining to the specification, design, manufacture, procurement, construction, inspection, assessment, management, operation, maintenance, refurbishment and demolition of buildings'. From this definition it is obvious that compliance or conformance can be physically observed during impromptu or planned inspections to achieve the goal of safety before, during and after construction activities – in other words, throughout all the phases of construction or the lifecycle of the building or structure.

It is therefore relatively easy for a construction organisation to identify non-compliance, non-conformance and defects and then introduce corrective or remedial interventions. It is for this reason that construction organisations focus most of their efforts on construction safety. Department of Labour inspectors may issue various notices for any perceived or observed contraventions of H&S legislation, regulations, safe work procedures, material safety data sheets, or codes of practice when they conduct inspections of any construction site. The leading causes of construction site fatalities continue to be falls from heights, electrocutions, crush injuries and caught-between injuries.

6.3 CONSTRUCTION HEALTH

Given the focus on construction safety, what is less recognised is that construction is also a high health risk industry. According to the Health and Safety Executive in the UK, every year more working days are lost due to work-related illness compared to injuries that result from poor construction safety. Construction workers have a high risk of developing diseases from several health issues. There are many different types of occupational disease including, for example:
- Respiratory diseases such as asthma and chronic obstructive pulmonary disease
- Skin diseases such as acute dermatitis

- Asbestos-related diseases such as asbestosis, mesothelioma and lung cancer
- Various cancers such as silicosis
- Noise-induced hearing loss
- Hand-arm and whole-body vibration syndrome
- Musculoskeletal disease
- Stress and depression

The most common health issues in construction according to the Health and Safety Executive (HSE) arise from the following:

- Cancers – past exposures in the construction sector cause over 5 000 occupational cancer cases and approximately 3 700 deaths every year. The most significant cause of these cancers is exposure to asbestos (70%), followed by extended exposure to silica (17%).
- Hazardous substances – dusts, chemicals and potentially harmful mixtures such as in paints are common in construction work. Some processes emit dusts, fumes, vapours or gases into the air, which in turn can be major causes of breathing problems and lung diseases. Several construction-related occupations also have high rates of dermatitis from skin exposures to hazardous substances, especially workers who regularly work with cement mortar, plaster and concrete.
- Physical health risks – skilled construction and building trades are occupations with the highest estimated prevalence of back injuries and upper limb disorders. Bricklaying is a particularly demanding construction activity. Manual handling is the most reported cause of lengthy injuries in the industry. Construction also has one of the highest rates of ill health caused by noise and vibration.

Occupational disease is a major issue the effects of which can be life-altering or life-ending. Traditionally, health issues in the workplace have been, and still are, harder to tackle than safety issues because their cause and effect are often not clearly linked. They are not usually visible and therefore treatable upon recognition. While some cases of ill health are clearly related to work activity, for others the cause may be less clear. Many serious occupational diseases such as lung cancer from asbestos exposure also have a long period of 'latency' of up to 30 to 40 years between exposure and development of ill health and/or disease, making the links even more difficult to establish. This phenomenon also means that after recognising the problem and making changes in working practices to reduce exposure there may be a long delay before a reduction in the causes of ill health and death are seen. However, where the link between work and occupational disease is established and exposure can be measured, interventions and activities aimed at raising awareness and creating behavioural change can work to reduce exposures and prevent ill health and disease.

Unfortunately, occupational health and disease in the construction industry are regarded as the poorer cousins of construction safety and therefore do not receive the attention that they need despite the severer impacts when compared with construction safety. The overall cost of treating occupational poor health and diseases is greater than that of rectifying poor safety. The only way to reduce losses and improve overall

H&S performance by construction organisations is to implement an effective H&S management system.

6.4 EFFECTIVE HEALTH AND SAFETY MANAGEMENT

Effective construction H&S management generally involves several elements, some of which are described briefly below.

A construction H&S policy statement must include a written policy document that has been signed by the CEO or most senior manager in the organisation. It is important that this policy document is widely displayed and contains a review date. This policy statement should be short enough and yet capture the essence of the approach of the organisation towards managing H&S effectively. A statement that fits on the back of a business card is recommended. Every worker should be able to recite this H&S policy statement.

The responsibilities of key persons or stakeholders must be documented in the management system, for example:
- Responsibilities of all construction workers
- Responsibilities of construction supervisors and foremen (OHS Act, section 8(2)(i) appointees)
- Responsibilities of senior management (OHS Act, section 16(1) and 16(2) appointees)
- Supervision of machinery (General Machinery Regulations (GMR) 2(1))
- Responsibilities of health and safety coordinators/representatives/officers
- Fire Fighter, Fire Fighting Coordinator, Fire Fighting Equipment Inspector
- Incident and accident investigation team
- Ladder Inspector
- Hazardous Chemical Substance Coordinator
- Ergonomics Survey Officer
- Pollution and environment survey team especially on green projects
- Work Permit Survey Officer, for example, when hot work is being done
- Air Power Tools Inspector and Explosive Power Tool Controller and operator when these are going to be used on the project
- Arrangements for construction H&S representatives and committees with reference to
  - Appointment of representatives per the GAR 6 and sections 17 and 18 of the Act
  - Establishment of health and safety committees per sections 19 and 20 of the Act
  - Appointment of Health & Safety Committee members
  - Appointment of Chairperson of Health & Safety Committee
  - Monthly meetings.

Several risk assessments should be done as required, namely:

- Baseline risk assessments are usually done prior to the commencement of any construction activity.
- Issue-based risk assessments focus on operational activities and processes and on the identification of the risks within a certain task, process or activity and are usually associated with the management of change.
- Continuous risk assessments are done on a consistent basis throughout each day to identify the critical hazards and exposures of current activities.

Any safe work or operating procedures that are known by the relevant or affected workers and physically available at each workstation are critical elements of any effective H&S management policy and plan. The following should be considered for inclusion:

- After the critical construction activities have been identified, the standards for these activities must be determined to establish safe and healthy work procedures for them.
- Written standard work or operating procedures (SWPs or SOPs) must be compiled for each of these critical activities to ensure uniform execution or implementation.
- Workers need to be properly trained to execute these construction activities strictly in accordance with the SWPs or SOPs.

Another important element of an effective construction H&S management system is that of worker orientation or induction, which will include training about construction H&S awareness on topics such as:

- Emergency procedures to be followed
- Nomination and introduction of first aider/s and the location of first aid stations
- Construction H&S responsibilities of the worker
- Details about any responsibilities as determined by general and construction H&S legislation and associated regulations
- Information about how injuries should be reported should these occur on construction sites or en route to them
- Explanations of unsafe conditions and acts relative to critical construction activities on projects
- Proper use and care of personal protective equipment as a measure of last resort after efforts have been made to mitigate the frequency and severity of any likely exposures
- Right to refuse hazardous construction work activities by workers
- Identifying hazards by supervisors and workers, including those outside their own work area
- Explanation of the reasons for and philosophy behind each construction health and safety rule
- Specific roles and responsibilities of each construction worker
- Scope of the authority of workers as determined by job descriptions that include details of when they have the right to stop work activities.

Details of construction H&S training programmes are a vital part of an effective H&S management system and should address some of the following elements:

- Knowledge and understanding of the construction H&S programmes, rules and procedures of the organisation as well as the specific roles and responsibilities of each construction worker

- Systematic programme of induction or orientation and ongoing training for construction workers and those who may be transferred between divisions of the organisation, jobs or tasks on construction sites
- Training of construction workers in the handling of any critical risks, hazards and dangers, precautions to be taken and procedures to be followed especially in emergencies
- Training in construction hazard identification, risk assessment, mitigation and control
- Training for all persons who may manage construction workers, contractors, co-contractors, suppliers and manufacturers
- Training of top or senior management in their important construction H&S roles and responsibilities
- Training and awareness programmes for contractors, co-contractors, temporary/casual workers, and visitors according to the level of risk that they could be exposed to on the construction site
- Training in the correct construction H&S inspection and reporting procedures
- Training in proper incident/accident investigation procedures
- Training in the effective monitoring of the quality and effectiveness of the construction H&S programme.

Inspections have been found to be one of the most effective ways to monitor construction H&S on construction projects. Therefore, an effective construction H&S management plan should include the following inspections:

- Those done by construction H&S representatives
- Those done by construction supervisors, foremen and leading hands
- Those prescribed by construction regulations and codes of practice
- Regular, planned construction project site inspections
- Frequent and routine construction equipment inspections
- Special inspections as required or as requested.

Note that standard construction project inspection lists could be useful tools. However, care should be taken to avoid checklist tick fever.

The following audits are necessary for effective H&S management:

- Internal audits
- External audits.

Given the need to effectively manage the health of construction workers, procedures for various medicals especially those that are a legal requirement must be put in place, for example:

- Procedures for pre-medicals, baseline, periodic and exit medicals where prescribed by a regulation or code of practice
- Baseline and regular periodic medicals where the construction site has exposure to

noise, heat, dust, chemicals, asbestos, lead, cannabis and coronaviruses
- Procedures for rapid testing, isolation and related requirements as prescribed for COVID-19.

The prescribed process of reporting, recording and investigating incidents and accidents must be strictly adhered to in the unfortunate event of these becoming necessary. The provisions and completion of documents required by section 24 (Reportable incidents (WC.I2)) and section 25 (Occupational diseases (WC.l1)) must be provided for.

It is critical that proper emergency procedures are developed and put in place. All workers must undergo training to be able to respond to emergency situations effectively. Regular drills need to be carried out from time to time. The following details need to be considered:
- Building construction details
- Access to and egress from all areas of the construction site
- Emergency exits on the construction site to designated safe areas that are clearly signposted

Figure 6.1: Example of emergency exit and accompanying instructions in Turkey

- Emergency lighting powered from an alternative source in the event of blackouts and load shedding
- Fire precautions and procedures
- Firefighting appliances that are sufficient, appropriate, and properly maintained and easily accessible
- Fire drills and alarm checks to ensure that the procedures are familiar to all workers on the construction site
- Proper storage of all flammables and combustible construction materials and consumables such as gas cylinders, fuels, oils, cleaning materials, wood and paper waste
- Identification of fire/explosion risk areas and instructions for isolating power and fuel should this become necessary
- Evacuation procedures and designated responsibility for roll calls
- Worker training in procedures and general fire safety and evacuation practices.

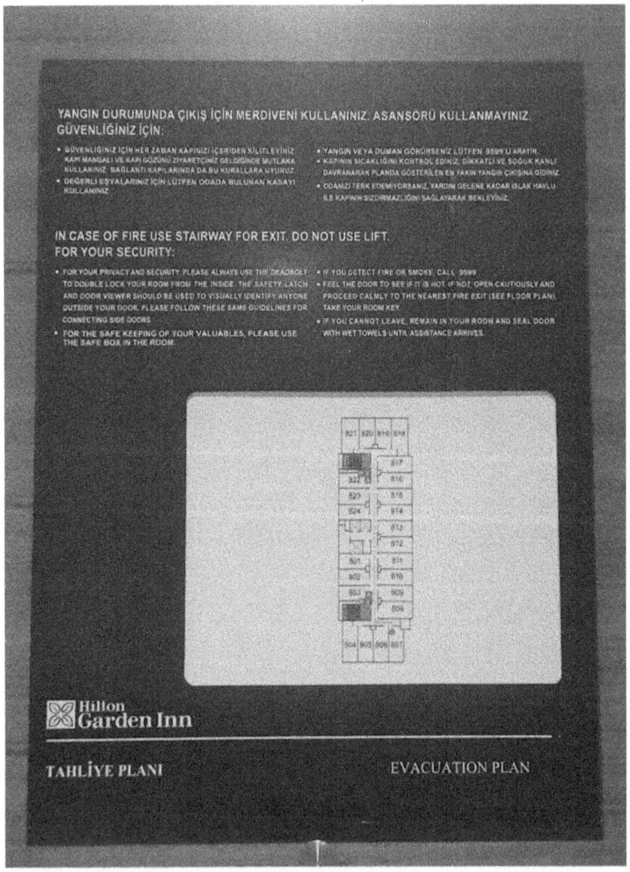

Figure 6.2: An example of emergency evacuation plan, procedures and explanatory guidelines in Konya, Turkey

Other emergencies and similar procedures to those for accidents and fire are required for emergencies to cover events, for example:
- Gas leaks
- Explosions
- Pressure vessel rupture
- Building collapse
- Floods, mud and landslides
- Chemical leaks and spillage
- Bomb threats
- Exposures to health threats such as outbreaks of COVID-19 infections.

The H&S policy and management plan should make provision for first aid with emphasis on the following:
- Proper training of first-aiders
- Provision of sufficient stocked first aid boxes, the contents of which should be carefully and regularly monitored
- Inspection and control of first aid boxes to ensure levels and adequacy of contents to identify and deal with any possible abuses.

How the organisation will manage contractors, co-contractors, suppliers and manufacturers must also be included in the H&S policy and management system and should cover the following:
- Contractors' agreements
- Validity of COID registration in the form of valid letters of good standing (LOGS)
- Medical certificates of all construction workers
- Current construction worker certifications and qualifications to demonstrate their competence to, for example, operate construction machines, plant and equipment
- Machinery and equipment checks, inspections and certifications
- Risk assessments for contractor and co-contractor activities.

Since large amounts of materials and potentially dangerous machinery are used on any construction project, procedures must be put in place to ensure that they are properly and safely handled and operated. Therefore, the following are important for effective H&S management:
- Handling of hazardous chemical substances and the provisions of Material Safety Data Sheets (MSDSs)
- Handling of hazardous biological agents
- Material handling rules
- Lockout and tagout procedures
- Hot work permits
- Maintenance programmes
- Vehicle safety rules
- Personal protective equipment requirements
- Engineering standards and procedures

- Purchasing and procurement standards and procedures
- Preventive maintenance.

6.5 OBTAINING MANAGEMENT COMMITMENT FOR A HEALTH AND SAFETY MANAGEMENT SYSTEM

- Any construction H&S management system can only be implemented with the commitment and involvement of the top or senior management of the construction organisation. This management commitment may be obtained by increasing the awareness of H&S throughout the entire organisation. Any proposal that contains the following information is most likely to get the commitment and involvement of senior management:
- Historic and current statistics and details of construction project-related injuries/incidents/deaths
- Inquiries from trade unions
- Pressure from construction project clients
- Financial losses due to poor construction H&S performance at organisation and project levels
- Moral and legal obligations towards construction workers to provide work environments that do not present a threat to their health and safety.

Other information that might also prove to be persuasive to management, includes the following:
- The current status of construction H&S in the organisation and on construction projects
- The number of construction workers injured in the past year
  - types of injuries sustained and numbers of each type
  - number of incidents on construction sites such as trips, slips and falls
  - financial loss incurred due to construction H&S-related incidents and accidents in the past year
  - number and size of fines and/or litigation especially actioned by the department of labour – current, pending and potential
- The proposed or desired status of construction H&S in the organisation and on construction projects with reference to:
  - health and safety policy and management system
  - number of injuries objective, which should preferably be zero but should at least show drastic gradual reduction over time
  - time frame within which to achieve this reduction in incidents and accidents and injuries that are sustained
  - extent of possible curbing, limiting or reducing financial losses from poor H&S performance.

6.6 CONSTRUCTION HEALTH AND SAFETY TRAINING AND AWARENESS

If improved construction H&S performance is to be achieved, it is well established both globally and nationally that relevant and regular H&S training and awareness are important. Construction worker involvement and engagement are necessary in those construction activities that directly affect them. Construction management must morally and legally provide a working environment on construction projects where workers are encouraged to make suggestions on working methods, equipment, materials, safety and health. These suggestions will enable management to take their ideas and opinions into account. Any management decisions that incorporate the suggestions from construction workers are more likely to be accepted by them on site than if they were just top-down.

A typical example could involve the design of a special scaffold to reduce pressure on the lower backs of bricklayers on site. If the ideas and thoughts of the bricklayers and their general workers are incorporated into the design of the scaffold, it is more likely that the introduction of the new scaffold will be accepted and used even if it involves a different approach to what has been the custom before. Undoubtedly, increased worker involvement and engagement build confidence and trust in the organisation, because it demonstrates that management actually regards construction workers and their views as important. As a result, the entire organisation senses that everyone is working together to achieve the shared and understood organisational objective of a safer and healthier working environment on construction projects.

When the views and opinions of construction workers irrespective of their standing in the organisation are taken into account, they generally are able to make substantive contributions in hazard identification, in proposing solutions and applying those solutions themselves. Therefore, H&S programmes are obviously more effective if workers have 'bought into them'. Their involvement and engagement must be more than just lip service. Rather, they must be shown to be meaningful and include decision-making capability as the construction H&S effort requires the participation and involvement of everyone in the organisation.

6.7 HEALTH AND SAFETY PUBLICITY AND PROMOTION

It is important that H&S publicity is part of the organisational education and training process within the H&S space. Construction H&S publicity and promotion may include public relations and low-key educational programmes designed specifically to inform rather than to persuade or change behaviour. For example, workers returning to work after the national lockdown due to COVID-19 will need to be informed about the 'new norm' in their workplaces since changes will have had to be made by the organisation to comply with legislated protocols that would affect them both directly and indirectly. H&S publicity and promotion programmes can be effective to promote and improve the H&S image of the organisation while at the same time being an effective tool to prevent accidents and incidents.

6.8 CONTRACTOR CONTROL

A contractor is generally defined as any person or party who comes on to a construction site or project with the explicit intention of performing any construction-related work. The Act specifically provides that the employer must ensure the H&S of all visitors on its premises or sites. These include other members of the project team such as architects, engineers and quantity surveyors who conduct meetings and progress inspections. Several incidents occur when these parties are not informed of the specific hazards on the construction site both generally and specifically during their time on the construction site or do not abide by the construction H&S rules and regulations on the specific site on which they are working or visiting.

Before any contractors who will be doing work on the site are allowed to enter any construction site it is important to verify the following:
•   The currency and validity of the status of workman's compensation registration with either the Compensation Commissioner or authorised private insurer such as Federated Employers Mutual Assurance
•   Details of the designated responsible person in charge of the work to be executed
•   Extent to which the construction workers who will be on site have been trained in construction H&S
•   Possession by all construction workers of the contractor of the necessary PPE who must be knowledgeable of their proper use, care and maintenance.

Clearly demarcate or mark all areas on the construction site where contractors are or will be performing their work and placed under the responsibility and supervision of the responsible contractor or its duly authorised representative on site. The primary contractor should ensure that work is performed according to all construction legislative and regulatory requirements, as well as the construction H&S rules of the organisation.

A contract should be signed by the contractor prior to the commencement of work on site which clearly indicates all the construction H&S requirements as well as the rules and regulations applicable to the work which is to be performed. The site-specific construction H&S plan of the contractor must be critically examined and approved in writing before any work is allowed to be carried out on the site. The contractors and their workers should also be inducted in the construction H&S rules of the organisation and the construction site, as well as the details of any risk assessments that are needed before any PPE is considered being used during the construction work to be performed.

All special requirements and arrangements that relate to the specific construction site and project must be explicitly explained to the contractor.

6.9 BUDGETING FOR CONSTRUCTION HEALTH AND SAFETY

It is better practice for all construction enterprises to have a specific budget that has been compiled to cover all the practical aspects of the construction policy of the organisation. A system of budgetary control ensures regular comparison of actual results with budget forecasts to ensure that the objectives of the construction H&S policy are met and form the basis for any required revisions of the policy.

A construction H&S manager must compile a H&S budget based on the overall construction H&S objectives of the organisation. This budget should consider the requirements of each construction project and at organisational level would comprise construction project H&S budgets based on the various risk assessments and mitigating interventions needed to respond to the anticipated exposure to significant residual project-specific hazards. Budgets include the following:

- A H&S budget prepared based upon the forecasts that highlight key task areas for action
- A production budget prepared in conjunction with these conditions and considering all the materials and other resources required to carry out those key H&S task areas
- The administrative cost budget for each area of activity
- A capital expenditure budget covering anticipated changes in legislation or purchase of specialised equipment or other modifications.

A construction H&S budget provides evidence that the organisation has given due consideration to the H&S aspects of each of their construction projects and their commitment to provide a safe and healthy working environment for all their construction workers, contractors and project stakeholders.

REVIEW QUESTIONS

1. Why is an organisational H&S policy necessary?
2. What strategy is necessary to convince the management of the organisation to introduce a H&S management programme?
3. What core elements should be included in a H&S management system?
4. Why is it necessary to have a documented H&S organisational structure?
5. What should the goals of a H&S propaganda programme be?
6. Why is a construction H&S budget necessary?

REFERENCES

Canadian Centre for Occupational Health and Safety. nd. *Effective Workplace Inspections.* [online] Ccohs.ca. Available at: https://www.ccohs.ca/oshanswers/prevention/effectiv. html

Department of Labour. 1993. Occupational Health and Safety Act 85 of 1993

Haupt, TC. 2021. *Management of Safety, Health and Environment in South Africa: A Handbook.* Newcastle upon Tyne: Cambridge Scholar Publishing

Health and Safety Authority (HSA). nd. *Safety and Health Management.* [online] Available at: https://www.hsa.ie/eng/Topics/Managing_Health_and_Safety/Safety_and_Health_Management_Systems/

Health and Safety Authority (HSA). nd. *Workplace Safety and Health Management.* [online] Available at: https://www.hsa.ie/eng/Publications_and_Forms/Publications/Safety_and_Health_Management/Workplace_Safety_and_Health_Management.pdf

International Labour Organisation. nd. *World Statistic.* [online] Ilo.org. Available at: https://www.ilo.org/moscow/areas-of-work/occupational-safety-and-health/WCMS_249278/lang—en/index.htm

International Organization for Standardization. 2015. *ISO14001 ENVIRONMENTAL MANAGEMENT SYSTEMS.* [online] Iso.org. Available at: https://www.iso.org/files/live/sites/isoorg/files/archive/pdf/en/pub100329.pdf

Lin, S, Mufidah, L & Persada, SF. 2017. Safety culture exploration in Taiwan's metal industries: Identifying the workers' background influence on safety climate. *Sustainability*, 9(11): 1965. Available at: https://doi.org/10.3390/su9111965

Occupational Safety and Health Administration. 2002. *Materials Handling and Storage.* [online] Osha.gov. Available at: https://www.osha.gov/Publications/osha2236.pdf

Occupational Health and Safety Assessment Series. 2007. OHSAS 18001:2007 Occupational Health and Safety Assessment Series. [online] Producao.ufrgs br. Available at: http://www.producao.ufrgs.br/arquivos/disciplinas/103_ohsas_18001_2007_ing.pdf

Occupational Safety and Health Administration. 2015. *OSHA Safety and Health Program Management Guidelines.* [online] Osha.gov. Available at: https://www.osha.gov/shpmguidelines/SHPM_guidelines.pdf

Safeopedia. What is the difference between a safety culture and a safety climate and why does it matter. Available at: https://www.safeopedia.com/what-is-the-difference-between-a-safety-culture-and-a-safety-climate-and-why-does-it-matter/2/8059

CHAPTER 7
MANAGING COMMON
CONSTRUCTION SAFETY HAZARDS

7.1 INTRODUCTION

The construction industry is complex both in nature and by its wide range of activities, which are executed in conditions that are unlike those in a factory where conditions can be controlled. Within the space of a single day it is possible for construction activities to be carried out under constantly changing conditions which present different safety hazards which have to be effectively managed to protect everyone involved. For example, outside activities are affected by changes in the weather ranging from working in heavy rain to high wind, from working in extreme heat to extreme cold and from working in dry to wet conditions. Workers could work on the ground floor, below ground level and also at various heights within the same shift. Multiple contractors could be working on the same site at the same time, thus presenting management challenges. Materials are sourced from multiple suppliers with coordination of deliveries and storage bringing their own challenges. In short, the construction industry is a hazardous and dangerous one, although not inherently so.

Several studies have identified the most common causes of accidents on construction sites. In some cases these accidents are caused by third-party negligence. The consequences of these exposures have resulted in injuries of various degrees of severity and fatalities. All construction employers have a duty as far as is reasonably practicable to safeguard their workers from these exposures. The most frequently cited causes of construction-related accidents include the following:

- Falls from heights such as when workers fall off ladders and scaffolding, leading to traumatic injuries and fatalities
- Falling objects where debris, materials and tools fall from scaffolding and higher levels
- Caught in between or being struck by objects, materials and equipment. Struck-by injuries are produced by forcible contact or impact between the injured person and an object or piece of equipment
- Struck-by hazards result in a person being squeezed, caught, crushed, pinched or compressed between two or more objects, or between parts of an object and can be classified as being struck by a falling object, being struck by flying objects, being struck by swinging objects, and being struck by rolling objects

Figure 7.1: Example of prevention from falling objects during construction activities during renovations in Australia

- Tripping hazards involving slips, trips and falls on the same or different levels, poor lighting, loose tools and greasy surfaces
- Defective equipment where tools have not been properly maintained
- Motor vehicle accidents, which are one of the most frequent incidents in South African construction
- Electrocution from contact with exposed wiring or incomplete electrical installations resulting in burns, electrocution, shock, arc flash/arc blast, fire and explosions
- Excessive noise from site and tools
- Vibration from tools and equipment used by construction workers
- Demolition accidents where explosives are used, for example, and result in burns, lacerations, fractured limbs, paralysis and fatalities
- Fires and explosions from gas and chemical leaks, electrical faults, incorrect handling of inflammable substances resulting in minor to serious degrees of burns, lung problems and fatalities
- Crane accidents due to possible lack of proper training, defective equipment and improper use and loading.

Figure 7.2: Noise, vibration and dust generated during deep excavation and demolition operations in Ankara, Turkey

Research has shown that the primary causes of fatalities in construction include unsafe methods, the nature of the industry, and the actual work environment. Other causes include working at height, incorrect and undocumented work procedures, and structural failure. However, the role of management upstream from the actual trigger event must not be overlooked because all accidents should be viewed as failures of management. Accident investigations invariably overlook the contribution and influence of management, who should not only be committed to but be visibly involved in construction health and safety management.

This chapter will address some of these causes and exposures with a view to the proactive prevention of the negative consequences of an accident occurring from them.

7.2 EXCAVATIONS

Most construction projects involve work below the natural ground level. Excavations whether shallow or deep are dangerous and hazardous activities. They may involve in the case of manual methods exposing construction workers to the risk of collapse or flooding resulting in cave-ins and entrapment of the affected workers, often with disastrous outcomes. In the case of mechanical methods the sides of deep excavations may collapse, spoil heaps placed too close to the edges of the excavations may slide into the excavations, and excavating machinery may capsize and fall into the excavation itself. Many construction workers are injured, maimed and killed as a result of excavation-related accidents. Where these accidents have occurred rescuers themselves have in some cases become victims.

Figure 7.3: Pipelaying in a deep excavation in Bahir Dar, Ethiopia, with supports to prevent risk of collapse

An excavation has been defined by the Occupational Safety and Health Administration (OSHA) in the USA as 'a man-made cut, cavity, or trench in the ground made by removing earth'. The most common type of excavation is the shallow trench, which is usually not wider than 5 metres.

Excavations can be executed safely provided a few principles are considered and procedures are strictly followed. While not an exhaustive list, the following are important considerations:

- Conduct on-site observations prior to commencing any work.
- Where geotechnical reports do not exist, take test cores and samples of the soil type and condition.
- Locate the presence of existing utilities and services that might be located underground.
- During excavation operations a competent person should conduct daily inspections of the work area and monitor any changing conditions that might affect the safety of the workers doing the excavation.
- Extra care should be taken after heavy rains or any other construction activities that might destabilise the excavations themselves.
- Provide or install warning systems that alert workers or equipment operators that they are working too close to the edges of excavations.

- Avoid construction workers working at levels above other workers in the excavation or underneath any loads handled by other workers.
- Protect workers in excavations by sloping or benching the sides of the excavations to acceptable angles of repose, supporting the sides of the excavations by means of shoring, and placing a shield between the side of the excavation and the work area in the form of a trench box.
- When installing excavation supports extra care must be taken to ensure that no part of the supports is overloaded and that another support is put in place before removing an existing one.
- Dismantling of support work should start at the bottom with each component removed carefully.
- After all components have been removed the excavation must be backfilled immediately and the backfill compacted.
- Adequate access and egress from particularly deep excavations must be provided for at regular intervals to allow workers to get into and out of excavations easily and quickly as needed.
- Workers must not be allowed to work in water which is accumulating unless proper dewatering processes are in place.
- Where excavations are deeper than 1.2 m the atmosphere in the excavation must be tested before work commences in that area.

Figure 7.4: Examples of deep trench excavations on a construction site in the Eastern Cape. Note the water and lack of side support

7.3 WORKING AT HEIGHTS

7.3.1 LADDERS

Ladders are used on almost every construction project irrespective of its size and project duration. Falls from heights involving ladders and stepladders make up a third of all reported fall-from-height incidents, resulting in many injuries and sadly also deaths. Many of these injuries are caused by inappropriate or incorrect use of the ladders.

Where work at height is necessary the use of a ladder or stepladder must be justified in terms of whether it is the most suitable access equipment to be used compared to other access equipment options. It is important that a risk assessment should be done and the following hierarchy of controls used:
* to avoid work at height where possible
* then to prevent falls from height
* then to reduce the consequences of a fall.

When considering whether it could be appropriate to use a ladder or stepladder, the following factors must be considered:
* Whether an alternative means of access is more suitable
* The nature of the work – ladders should only be used for 'light work' as they are not suitable for strenuous or heavy work
* The duration of the work activity is preferably not for longer than 30 minutes
* The height to be worked at
* What reaching movements may be required while on the ladder as workers must not overreach
* What materials and equipment may be required to be used at height, noting that where the task involves a construction worker carrying more than 10 kg up the ladder or steps the activity will need to be justified by a detailed manual handling assessment
* Whether it is possible to maintain three points of contact (hands and feet) at the working location and where not possible, other measures must be considered to prevent a fall or reduce the consequences of one
* Where a proper handhold is not practicable or possible, a risk assessment will have to justify whether the use of a ladder is safe or not
* Whether it is likely that the ladder or stepladder will be overloaded if the person and anything they are carrying will exceed the highest load stated on the ladder
* Whether the work to be done on the ladder or stepladder will impose a side loading as stepladders are not designed to accept any side loading.

The Occupational Health and Safety Act 85 of 1993 and General Safety Regulation 13A (Ladders) stipulate as follows:

(1) A worker shall ensure that every ladder is constructed of sound material and is suitable for the purpose for which it is used, and –

    (a) is fitted with non-skid devices at the bottom ends and hooks or similar devices at the upper ends of the stiles which shall ensure the stability of the ladder during normal use; or

    (b) is so lashed, held or secured whilst being used as to ensure the stability of the ladder under all conditions and at all times.

(2) No employer shall use a ladder, or permit it to be used, if it –

    (*a*)   (i)   has rungs fastened to the stiles only by means of nails, screws, spikes or in like manner; or

           (ii) has rungs which have not been properly let into the stiles: Provided that in the case of welded ladder   or ladders of which the rungs are bolted or riveted to the stiles, the rungs need not be let into the sides; or

    (*b*) has damaged stiles, or damaged or missing rungs.

(3) No employer may permit that –

    (*a*) a ladder which is required to be leaned against an object for support be used which is longer than 9 m; and

    (*b*) except with the approval of an inspector, the reach of a ladder be extended by fastening together two or more ladders: Provided that the provisions of this sub-regulation shall not apply to extension of free-standing ladders.

(4) In the case of wooden ladders the employer shall ensure that –

    (*a*) the ladders are constructed of straight grained wood, free from defects, and with the grain running in the length of the stiles and rungs; and

    (*b*) the ladders are not painted or covered in any manner, unless it has been established that there are no cracks or other inherent weaknesses: Provided that ladders may be treated with oil or covered with clear varnish or wood preservative.

(5) When work is done from a ladder, the employer shall –

    (*a*) take special precautionary measures to prevent articles from falling off; and

    (*b*) provide suitable sheaths or receptacles in which hand tools shall be kept when not being used.

(6) An employer shall ensure that a fixed ladder which exceeds 5 m in length and is attached to a vertical structure with an inclination to the horizontal level of 75° or more –

    (*a*) has its rungs at least 150 mm away from the structure to which the ladder is attached; and

    (*b*) is provided with a cage which –

           (i)   extends from a point not exceeding 2.5 m from the lower level to a height of at least 900 mm above the top level served by the ladder; and

           (ii) shall afford firm support along its whole length for the back of the person climbing the ladder, and for which purpose no part of the cage shall be more than 700 mm away from the level of the rungs:

Provided that the foregoing provisions of paragraph (b) shall not apply if platforms, which are spaced not more than 8 m apart and suitable for persons to rest on, are provided.

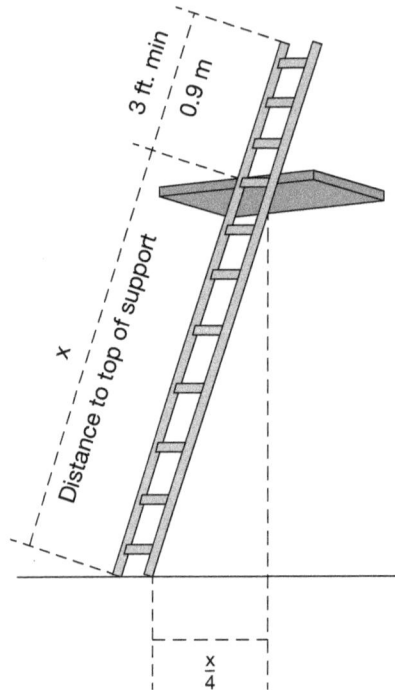

Figure 7.5: Placement of ladder against a wall

- Always avoid bringing the ladder into contact with electricity or within 6 m horizontally of any overhead power lines.
- Always use a non-conductive ladder or steps for any necessary live electrical work.
- Avoid using ladders or stepladders where they could be struck by vehicles; this can be prevented by protecting them with suitable barriers or visible plastic cones.
- Always avoid using ladders or stepladders where they could be pushed over by other hazards such as doors or windows – secure doors (not fire exits) and windows where possible. If this is impractical, have a person standing guard at a doorway, or inform workers not to open windows until they are told to do so.
- Do not use ladders or stepladders where pedestrians could walk under them or near them – use barriers, cones or, as a last resort, a person standing guard at the base.
- Place the ladder so that its feet are a quarter of its length to the top support from the object it is resting against.
- Unless the ladder is securely tied at the top it should always be held in position by another person while in use.
- Wherever possible it should extend 1 m above its support.

Figure 7.6: This ladder is not securely tied at the top and does not extend 1 m above its support, during a window cleaning operation in Cape Town

- Ladders should be inspected at frequent intervals. Only use ladders or stepladders that have no visible defects. They should have a pre-use check each working day in accordance with the manufacturer's instructions. Ladders that are part of a scaffold system must be inspected every seven days. Ladder and stepladder feet must be part of the pre-use check. Defective ladders should be repaired or replaced.
- As paint conceals defects, use varnish or two coats of oil instead, to preserve wooden ladders.
- Keep ladders clean because dirt hides defects while grease or oil cause people to slip.
- Do not leave ladders lying on wet ground or exposed to the weather.
- Ladders lying on floors may cause workers to trip and fall or may be run over by vehicles and be damaged.
- Do not use ladders horizontally as walkways, or scaffolding.
- Tools and equipment should be hauled up by ropes.
- Never leave a ladder where it may fall.
- Never place a ladder in front of a doorway before taking adequate precautions.
- Ladders should be equipped with safety feet. The feet should be in good repair, secure and clean, and not loose, missing, splitting or excessively worn. The feet should be in contact with the ground.

- Check ladder feet when moving from soft or dirty ground such as dug soil, loose sand/stone or a dirty workshop to a smooth, solid surface such as paving slabs, to ensure the foot material and not the dirt such as soil or embedded stones is making contact with the ground.
- Store ladders in a cool place, either lying on their sides or hanging in a horizontal position from several wall brackets. Avoid warping wherever possible.
- Use both hands when climbing up or down a ladder.
- Only one person at a time should use a ladder.
- Do not leave tools or other equipment on the top of a ladder. Remove these items when descending from a ladder.
- Use the correct ladder for the job.

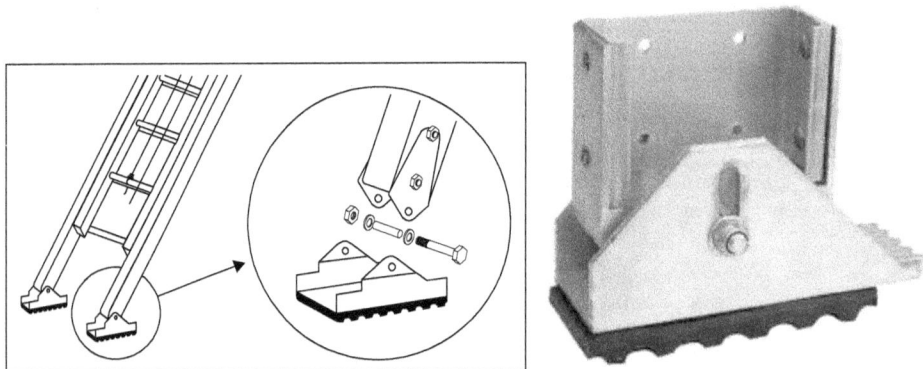

Figure 7.7: Example of ladder feet

7.3.2 SCAFFOLDING

Working on scaffolding is one of the most dangerous construction activities. Most accidents involving scaffolding are caused by one or a combination of planking failure, support failure, slipping or the impact of a falling object. A scaffold is defined as a type of temporary elevated platform and their structure that are used for supporting workers and materials at a distance above the ground. The definition does not prescribe the type of material to be used to form the scaffold. In many countries because of the non-availability and prohibitive cost of metal scaffolding systems timber and bamboo have been used since time immemorial.

Figure 7.8: Example of timber scaffolding on a construction renovation project in Ethiopia. Note how the members are tied together with twine

The safe use of scaffolding demands attention to several key aspects, which include the following:

- Ensure that the scaffold can support its own weight plus an additional load of at least four times the maximum intended load.
- In the case of suspension scaffolds, the ropes must support the weight of the scaffold itself plus six times the maximum intended load.

Every platform of the scaffold must be fully planked and generally be at least 450 mm wide.

Figure 7.9: Metal scaffolding system used during renovations to the Blue Mosque in Istanbul, Turkey. Note the platforms, bracing and supports

- Where long scaffolds are required like across the face of an entire building, platforms may be placed end to end with each platform resting on its own supports.
- The use of scaffold components made from different materials must be avoided unless approved by a competent person.
- Ensure that the 4:1 rule is adhered to where the height of a freestanding scaffold must not exceed four times the width of the base of the scaffold. Where the 4:1 rule is exceeded, special supports must be introduced.
- Scaffolds must always be placed on a firm footing.

Figure 7.10: An example of a scaffold support on a poor timber footing and an example of the use of screening to protect workers and the public against falling objects

- The scaffolding must be inspected before the beginning of any work shift.
- Scaffolds must not be allowed to come into contact with energised electrical powerlines with a general distance of more than 3 m.
- Workers must be protected against falling from scaffolds by personal fall arrest systems and/or guard rails.
- Fall arrest systems must be attached to a vertical or horizontal lifeline or a structural component of the scaffold by means of a lanyard.
- Guard rails must be installed on all open sides of a scaffold platform as well as the scaffold ends.
- To protect against falling objects, construction workers must always wear hard hats, with the scaffold being fitted with toe-boards, screens, and catch platforms.

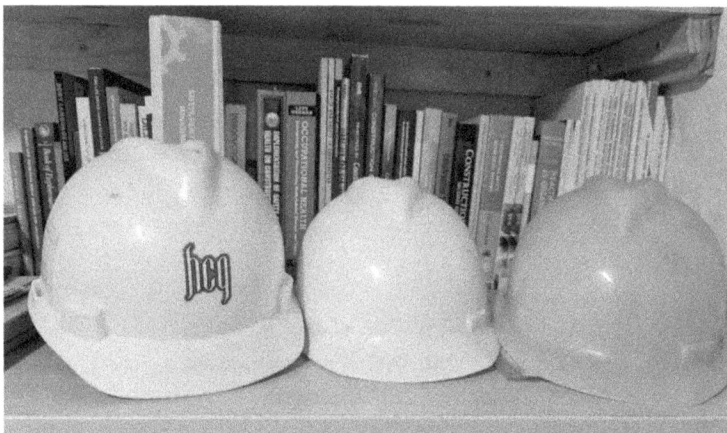

Figure 7.11: Wear a hard hat, which comes in different sizes and colours, for head protection

REGULATIONS CONCERNING SCAFFOLDING

The following regulations cover the safe design, construction and use of scaffolding.

Regulation 13D (Scaffold framework)

(1) An employer shall ensure that –

    (a) Scaffold standards are properly propped against displacement and are secured vertically on firm foundations: Provided that putlog scaffolds shall incline slightly towards the structure;

    (b) (i) Steel scaffold standards with 'heavy', 'medium', 'light' or 'very light' platform loadings which shall not exceed 320, 240, 160 and 80 kg/m², respectively, are spaced not more than 1.8 m, 2 m, 2.5 m and 3 m apart, respectively; and

        (ii) wooden scaffold standards are spaced not more than 3 m apart;

    (c) Ladders are spaced vertically not more than 2.1 m apart;

    (d) Putlogs or transoms –

        (i) which do not support a platform, are spaced at the same distances as the distances prescribed in paragraph (b) in respect of scaffold standards;

        (ii) which support a platform, are spaced not more than 1.25 m apart if the platform is constructed of solid timber boards;

    (e) every part of a wooden scaffold frame has a diameter of at least 75 mm or a section of similar strength.

(2) No employer shall use a scaffold, or permit it to be used unless it –

    (a) is securely and effectively braced to ensure stability in all directions;

    (b) is secured at suitable vertical and horizontal distances to the structure to which work is being done, unless it is designed to be completely freestanding;

    (c) is so constructed that it has a throughout factor of safety of at least two; and

    (d) is inspected at least once a week and every time after bad weather by a person who has adequate experience in the erection and maintenance of scaffolds, and all findings are recorded in a register or report book.

(3) No employer shall require or permit that –

    (a) a scaffold with a supporting wooden framework exceeds a height of 10 m; and

    (b) a scaffold is erected, altered or dismantled by or under the supervision of a person other than a person who has had the necessary training and experience of such work and who has been appointed by the employer in writing for this purpose.

Regulation 13E (Scaffold platforms)

(1) An employer shall ensure that –

    (a) every plank of a solid wooden scaffold platform is at least 225 mm wide and 38 mm thick;

    (b) every plank which forms part of a scaffold platform is supported at distances not exceeding 1.25 m, and its ends are projected not less than 70 mm and not more than 200 mm beyond the last prop;

    (c) every plank of a scaffold platform is firmly secured to prevent its displacement; and

    (d) every platform is so constructed as to prevent materials and tools from falling through.

(2) An employer shall ensure that every scaffold platform –

    (a) with 'heavy', 'medium', 'light', or 'very light' platform loadings as referred to in regulation 13D (1) (b) (i) is not less than 1 125 mm and not more than 1 380 mm, not less than 1 125 mm and not more than 1 150 mm, not less than 900 mm and not more than 1 150 mm, and not less than 675 mm and not more than 1 150 mm, respectively, wide: Provided that where a platform is used only as a gangway, a platform width of 450 mm shall be sufficient;

    (b) which is more than 2 m above the ground is on all sides, except the one facing the structure, provided with –

        (i) substantial guard rails of at least 900 mm and not exceeding 1 000 mm in height; and

        (ii) toe-boards which are at least 150 mm high from the level of the scaffold platform and so affixed that no open space exists between the toe-boards and the scaffold platform: Provided that the toeboards are constructed of timber, they shall be at least 25 mm thick;

    (c) is not more than 75 mm from the structure: Provided that where workmen must sit to work, this distance may be increased to not more than 300 mm; and

    (d) is kept free of waste, projecting nails or any other obstructions, and is kept in a nonslip state.

(3) No employer shall require or permit that a working platform which is higher than 600 mm be supported on a scaffold platform and shall provide an additional guard rail of a least 900 mm and not exceeding 1 000 mm in height above every such working platform.

(4) An employer shall ensure that convenient and safe access is provided to every scaffold platform, and where the access is a ladder, the ladder shall project at least 900 mm beyond the top of the platform.

Figure 7.12: Example of toe-boards on scaffolding

Note: To comply with regulation 13D(3)(*b*) an employer must appoint in writing a supervisor, who has had the necessary training and experience, to supervise the erection, maintenance, dismantling and inspection of all scaffolding.

• To comply with regulation 13D(2)(*d*) a scaffold inspection register must be kept.

TYPICAL COMPONENTS OF A SCAFFOLD

Depending on the scaffolding system being used, a scaffold typically consists of the following components:

• Sole plate, which is an essential component that spreads the weight of the scaffold over a greater area. While usually made of timber they may also be made from metal or concrete.
• Adjustable base plates from which different height-adjustable base plates can be selected that come with strong and self-cleaning round threads to adjust to the ground. They can have colour and notch markings to render safeguard against over-winding.
• Standards, which are also known as uprights and are perpendicular tubes that shift the entire weight of the structure to the ground where they lean on a square base plate to distribute the weight.
• Ledgers, which are flat tubes that join between the standards.
• Transoms, which lean on the ledgers at right angles. Major transoms are positioned next to the standards; they support the standards that are in place and give support for the boards. To render additional support for the boards, intermediate transoms are placed between the main transoms.

- Diagonal bracing, which further supports the basic structure comprising vertical standards and ledgers. Additionally, their high connection standards assist special structures.
- Decking, which are also referred to as planks and available, for example, in aluminium, aluminium frame with plywood board and hot-dip galvanised steel.
- Guard rails (Handrails and mid-rails) height approximately 1 metre.
- Toe-boards, which are placed between vertical standards and are obtainable in aluminium, steel or wood. The steel toe-board lowers the fire hazard and lasts long. Because of its design, there are no openings or gaps between the deck and the toe-board. They should be minimum 150 mm in height.
- Ladders, which should be secured at three locations within the scaffold.

Figure 7.13 shows an example of a basic putlog scaffold with the various components.

Figure 7.13: Example of basic scaffolding in Sydney, Australia

Figure 7.14: Note the boarding, toe-boards, bracing, and partial screening in a metal scaffolding system on a construction site in Sydney, Australia

SCAFFOLDING HAZARDS

Apart from collapse of the scaffolding, the main hazards for construction workers working on scaffolds are the following:
- Falls from height, which are caused by slipping, unsafe access and the lack of fall protection
- Being struck by falling materials, tools and rubble or debris
- Electrocution from, for example, the scaffold touching energised overhead electrical power lines
- Slipping of an unsecured ladder within the scaffold
- Use of unsuitable, damaged and faulty materials in building the scaffold
- Inadequately supported scaffold boards or platforms
- Omission and/or removal of guard rails and toe-boards
- Scaffolding not being properly braced
- Overloading of platform and board beyond the maximum intended load.

Figure 7.15: Metal scaffolding system around a minaret in Istanbul, Turkey. Note the working platforms and high level of bracing

Falls from scaffolds happen:
- while workers are climbing on or off the scaffold
- when workers trip over obstacles on the scaffold platform such as debris, material and equipment
- while workers work on unguarded scaffold platforms
- when workers slip on the surface of the platform or scaffold boards
- when scaffold platforms or planks fail
- when workers are not properly tied off
- when workers incorrectly dismantle the scaffold after use.

Table 7.1: Fall protection required for specific scaffold types

Type of scaffold	Guard rails required	Personal fall arrest system required	Both required
Aerial lifts		✔	
Boatswains' chair		✔	
Catenary scaffold		✔	
Chicken ladder*	✔	✔	
Float scaffold		✔	
Ladder jack scaffold		✔	
Needle neam scaffold		✔	
Self-contained scaffold			✔
Single-point and two-point suspension scaffolds			✔
Supported scaffold	✔	✔	
All other scaffold	✔	✔	

*Chicken ladders require guard rails or personal fall arrest system or a grab-line system

(Source: https://www.constructionsafety.co.za/special-areas/heights-scaffolding/)

Falls from scaffolds can be prevented by:
- keeping the construction site free of obstacles and debris
- keeping the surface of platforms and boards clean and dry
- avoiding running cables or cords on the walkways that could become tripping hazards
- using ladders in the scaffold correctly
- ensuring that scaffolds are erected by a competent person and follow a clearly visible tagging system
- wearing proper safety footwear
- maintaining adequate lighting where a scaffold is used indoors or in a poorly lit area
- using a proper safety harness or fall arrest system if working at height (more than 1.8 m) and always be tied off correctly.

Figure 7.16: Examples of harnesses and being tied off during the erection of wooden electric poles in the Eastern Cape

It is critically important to note that scaffolds should be used as means of access to elevated areas of work and as working platforms. They are not support work. Where failures occur, large areas of scaffolding can collapse quite suddenly. Scaffolds can collapse because of poor construction or misuse, leading to them being loaded beyond their safe capacity to support the load. Common faults are poor foundations, inadequate tying and bracing, overloading and the removal of ties and bracing.

Scaffold stability depends on carefully following the scaffold system manufacturer's instructions and the provisions of health and safety codes and other equivalent standards.

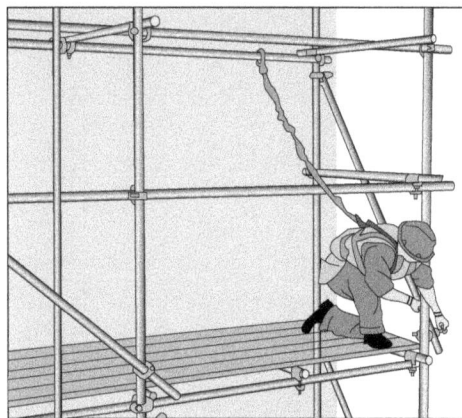

Figure 7.17: Example of being properly tied off while working on a scaffold

LOADING BAYS IN SCAFFOLDS

A properly constructed loading bay can avoid the excessive loading of access scaffolds and the obstruction of gangways that can otherwise occur. Loading bays are traditionally designed for a 'blanket' loading of 1 tonne per square metre (10 kN/m^2), plus an allowance of 25% for light mechanical loading of materials.

Figure 7.18: Example of support work on a site in Hong Kong

In terms of their design, loading bays should be diagonally braced on all four sides or braced in compliance with the scaffold system manufacturer's recommendations. Where the internal facade bracing hinders access onto the scaffold from the loading bay, the brace may be placed on the main scaffold adjacent to the loading bay or in accordance with the system manufacturer's recommendations. The following must be considered:

- Standard transoms at standard spacing and timber decking at standard spans are not usually adequate to carry the higher loadings in a loading bay. Certain systems of scaffold loading bays incorporate special load-bearing transoms, often at reduced spacing.
- Where load-bearing transoms are directly connected to the outside face of a scaffold, the capacity of the standards to support the combined loads imposed by the working platforms and the load-bearing transoms should be assessed.

- Plan bracing should be installed from the outside corner of a loading bay to the main access scaffold and the main scaffold should be tied to the building with supplementary ties opposite these braces at intervals not exceeding 3 m.
- Where guard rails must be removed temporarily to facilitate loading, effective compensatory measures to prevent falls should be provided. These measures may include movable guard rails or panels, hand holds or harnesses affording an equivalent standard of protection as guard rails. Temporarily unguarded openings or edges should not be left unattended and guard rails should be replaced as soon as practicable.

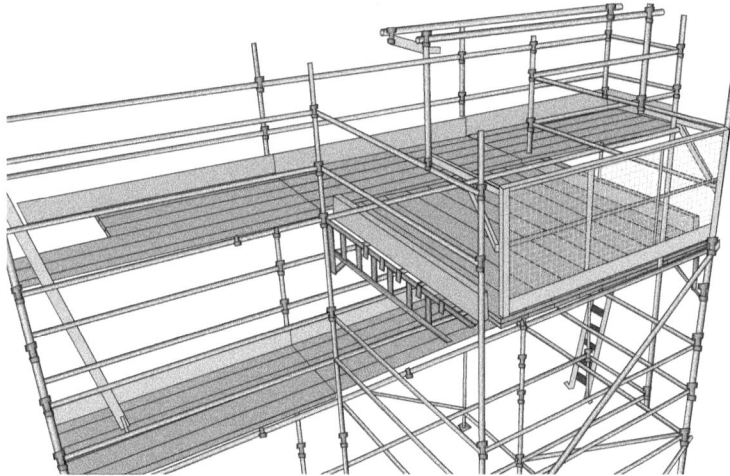

Figure 7.19: An example of the proper design of loading bay

(Source: https://accesspoint.org.uk/technical-loading-bay-design/)

- Easily comprehensible signs showing the safe working load should be placed on scaffolds and loading bays.

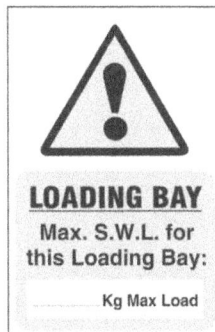

Figure 7.20: Example of safe loading sign

(Source: https://www.pdsigns.ie/product/safety-construction-hazard-max-load-for-this-loading-bay-sign/ – Used with permission)

7.3.3 CRANES

Modern-day construction work takes place in a highly mechanised working environment in which there is a heavy reliance on mechanical material handling and lifting. Therefore, cranes have become an essential part of the construction process and are widely used to move materials around the construction site, which can help resolve some of the health and safety problems arising from space constraints and workplace transport issues. The growing industrialisation of construction processes and off-site manufacture of often large and heavy modular components have made cranes a critical feature of construction operations. However, cranes are regarded as being one of the most dangerous items of equipment on a construction site. Crane safety is a concern in construction operations across the globe.

The type of crane to be used on the construction project will depend on several factors such as the size of the load, the location of the loading and unloading site, the type of load, the distance and height the load must be moved and the conditions under which the load must be moved. To choose the correct crane for the project, the difference must be known between the different types of cranes. There are specific legal requirements to ensure that all cranes are installed, inspected, examined and maintained to ensure the risks relating to lifting are adequately controlled. Several investigations into recent crane-related accidents have shown that enhanced standards of planning, management and execution of the crane erection and operation processes could have reduced the chance of death and injury.

Figure 7.21: Examples of different types of cranes

(Source: https://www.velvetcushion.com/other/overhead-crane-safety-101)

Contractors must only allow thoroughly trained, competent and certified crane operators to operate cranes. Operators should know what they are lifting and what it weighs. For example, the rated capacity of mobile cranes varies with the length of the boom and the radius. When a crane has a telescoping boom, it is extended and the radius increases. The following steps should be taken to improve crane safety:

- Ensure that all cranes are erected and dismantled by competent people who have the necessary crane-related training and experience. Construction companies need to have written procedures for each type of crane which should be based on the instructions of manufacturers. These procedures should be available where the crane is being used and those involved in the work must be familiar with them.
- Ensure that a thorough examination of the crane is undertaken after its erection by a competent person who is sufficiently independent and impartial and is not involved in the erection process.
- Only allow competent people with the necessary crane-related training and experience to operate the crane.
- Ensure that the crane operator carries out pre-use checks at the start of each work shift to ensure that the crane has not suffered any damage or failure during the previous shift and is safe to be used.
- Ensure that all cranes that have adjustable booms are equipped with boom angle indicators.
- Provide cranes with telescoping booms with some means to determine boom lengths unless the load rating is independent of the boom length.
- Post load rating charts in the cab of cab-operated cranes.
- Require workers to always check the crane's load chart to ensure that the crane will not be overloaded by operating conditions.
- Instruct construction workers to plan crane lifts before starting them to ensure that they are safely executed.
- Inform workers to take additional precautions and exercise extra care when operating in proximity to energised electrical power lines to prevent accidental electrocutions through contact.
- Train workers to always ensure that outriggers on mobile cranes rest on firm ground, on timbers, or are sufficiently cribbed to spread the weight of the crane and the load over a large enough area on the site.
- Direct workers to always keep hoisting chains and ropes free of kinks or twists and never wrapped around a load.
- Train workers to attach loads to the load hook slings, fixtures and other devices that have the capacity to support the load on the hook.
- Instruct workers to properly pad sharp edges of loads to prevent cutting slings
- Teach workers to maintain proper sling angles so that slings are not loaded in excess of their capacity.
- Ensure that all cranes are inspected frequently by persons thoroughly familiar with the type of crane, the methods of inspecting the crane and what can make the crane unserviceable. Inspection schedules should be determined by the expected crane

activity, the severity of its use and the prevailing environmental conditions. In-service inspections are carried out by the crane operator, generally at weekly intervals, and records kept of these inspections.

- Ensure the critical parts of the crane such as the crane operating mechanisms, hooks, air or hydraulic system components, and other load carrying components are inspected daily to identify any possible maladjustment, deterioration, leakage, deformation or other damage.

- A properly planned maintenance system must be established and adhered to. Competent people should undertake this maintenance at the intervals specified by the manufacturer and records kept of the work completed including any parts that have been replaced. In general, the original manufacturers parts should always be used. Where parts are sourced from suppliers other than the original manufacturer, a competent engineer should assess that the parts selected meet the original manufacturer's specifications and are fit for purpose. Any parts replaced should be installed in accordance with the instructions of the manufacturer.

Figure 7.22: An example of a mobile crane capsizing because of poor founding of the outriggers and overextension of the telescopic boom

IMPORTANCE OF LOADING CHARTS

Supervisors and equipment operators such as crane and telescopic fork truck drivers, must be provided with easily comprehensible loading charts showing the weights of the typical materials used on the construction site. This chart or list should include, for example, the weights of the pallets of bricks and blocks, scaffold boards and standards, and mortar skips. In this way construction workers will be able to estimate the load they are placing on the crane and ensure that it is less than the safe working load indicated on the signs.

BOOM LENGTH												Maximum Load Chart in pounds (lbs) with fully extended outrigger
	27 FT		34 FT		43 FT		52 FT		61 FT		70 FT	
OPERATING RADIUS (FT)	LOADED BOOM ANGLE (DEG)	LOAD RATING (LB)	LOADED BOOM ANGLE (DEG)	LOAD RATING (LB)	LOADED BOOM ANGLE (DEG)	LOAD RATING (LB)	LOADED BOOM ANGLE (DEG)	LOAD RATING (LB)	LOADED BOOM ANGLE (DEG)	LOAD RATING (LB)	LOADED BOOM ANGLE (DEG)	LOAD RATING (LB)
5	77	34,000*										
10	66	21,100*	71	17,100*	75	16,000*	78	15,700*				
15	54	15,100*	62	14,000*	68	12,100*	72	11,100*	75	10,800*	77	9,600*
20	39	11,100*	51	10,100*	61	9,100*	66	8,600*	71	8,200*	73	7,300*
25	17	7,900*	40	7,700*	53	7,100*	61	6,900*	66	6,600*	69	5,900*
30			23	6,500*	44	6,100*	54	5,600*	60	5,300*	64	4,900*
35					33	4,800*	47	4,700*	54	4,660*	60	4,150*
40					16	3,500*	38	4,100*	48	3,900*	55	3,550*
45							27	3,250*	41	3,200*	49	3,050*
50							9	2,950*	33	2,800*	44	2,650*
55									23	2,500*	37	2,350*
60											29	1,900*
65											19	1,700*
STOWED JIB DEDUCTIONS (POUNDS)												
	450		360		260		230		200		175	

NOTE: STRUCTURAL STRENGTH RATINGS IN CHART ARE INDICATED WITH AN ASTERISK *

Figure 7.23: Example of loading chart for cranes

(Source: https://www.cranehunter.com/how-to-read-crane-load-chart)

7.3.4 HOISTS

In the construction industry hoists are generally classified as material and personnel or passenger hoists. Each type of hoist has specific requirements and safety practices. Important information must be posted in clear and unobstructed places on cars and platforms. This information includes the rated load capacities, recommended operating speeds, special hazard warnings and any other special instructions, for example, wearing of masks and keeping of minimum distances of 1,5 m and maximum number of workers allowed in the cars and platforms to comply with COVID-19 protocols. Ensuring the wire ropes used with material hoists are properly installed and capable of carrying the expected maximum load is critical for the safety of everyone on the site. Openings to hoists must be guarded by substantial gates or bars that cover the entire width of the opening. Overhead protection must also be provided on the top of all material hoists and platforms. Car arrest systems must be installed to function automatically in the event of rope failure. Towers for passenger hoists must be enclosed on all sides and must be anchored to the structure every 8 m. The cars must be enclosed on all sides, including the top. Doors and gates must be fitted with mechanical locks that cannot be operated from the landing side.

Figure 7.24: Example of a material and passenger hoist in Melbourne, Australia

7.4 HOUSEKEEPING

Incidents involving slips, trips and falls on the same level are growing concerns in the construction industry with substantial injury and economic consequences. These often are caused by poor housekeeping, slippery floors, obstacles on the path and poor or uneven floor/pavement surfaces on construction sites. Housekeeping in the context of the construction industry is defined by SANS 10091:2015 as the use of those procedures designed to bring about and maintain cleanliness, tidiness and orderliness while working on a construction site or workplace.

A place for everything and everything in its place on the construction site or project

On a construction site, good housekeeping will involve, for example:
- Clean hygienic and properly sanitised ablution facilities
- Good stacking of and storage practices for materials such as bricks, timber and roof sheets
- Tidy construction plant yards
- All refuse being kept in clearly marked dumpsters on the construction site
- Clear and uncluttered scaffold platforms.

On the other hand, poor housekeeping on the construction site would involve, for example:
- Unnecessary rubbish, waste and vegetation accumulation on the site
- Poor and untidy stacking and storage practices
- Unsafe handling of flammable liquids.

Effective housekeeping would therefore require the following:
- All working areas and surfaces on the site should be kept clear of any obstacles to prevent workers from injuries because of slips, trips and/or falls.
- Dumpster bins should be provided in strategic locations on the site to collect any waste materials and debris that need to be removed regularly from the site.
- All spillages on the construction site must be cleared and cleaned immediately to prevent workers from slipping or being contaminated if chemicals are involved.
- All areas on the site that were worked in during the day should be cleaned up at the end of every day.

Figure 7.25: An example of poor housekeeping in Ethiopia. Note the upward facing, protruding nails

All workers should have a working knowledge of the following:
- The meaning of any warning and prohibitory signs
- The location of firefighting equipment such as extinguishers and hoses
- Who the health and safety representative and first-aider on the site are and where they can be found
- The locations of clearly and visibly marked emergency exits.

If slips, trips and falls on the construction site are to be avoided, the following should be noted:
- Walking under suspended loads from cranes must be avoided
- Increased awareness of any construction workers working at height or overhead
- Watching out for any moving motor vehicles on site, as in South Africa motor vehicle accidents dominate as causes of workplace accidents and injuries
- Watching out for any moving plant and equipment on the site.

Figure 7.26: An example of good housekeeping on a site in Hong Kong

7.5 STACKING AND STORAGE

If stacked incorrectly on the construction site, materials and other supplies can fall and cause injuries like cuts and bruises or even more serious injuries related to crushing and pinning. Contractors need to ensure that their workers follow a set of standards for the storage of materials on site to avoid these accidents.

7.5.1 GENERAL SAFETY REGULATION 8

(1) No employer shall require or permit the building of stacks which consist of successive tiers, one on top of another, unless –
  (a) the stacking operation is executed by or under the personal supervision of a person with specific knowledge and experience of this type of work;
  (b) the base is level and capable of sustaining the weight exerted on it by the stack;
  (c) the articles in the lower tiers are capable of sustaining the weight exerted on them by the articles stacked above them;
  (d) all the articles which make up any single tier are consistently of the same size, shape and mass;
  (e) pallets and containers are in good condition; and
  (f) any support structure used for the stacking of articles is structurally sound and can support the articles to be stacked on it.

(2) An employer shall not permit –
  (a) articles to be removed from a stack except from the top most tier or part of that tier; and
  (b) anybody to climb onto or from a stack, except if the stack is stable and the climbing is done with the aid of a ladder or other safe facility or means.

(3) An employer shall take steps to ensure that –
  (a) persons engaged in stacking operations do not come within reach of machinery which may endanger their safety;
  (b) stacks that are in danger of collapsing are dismantled immediately in a safe manner; and
  (c) the stability of stacks is not endangered by vehicles or other machinery or persons moving past them.

(4) Unless a stack is otherwise supported, an employer shall take steps to ensure that tiers of stacked material consisting of sacks, cases cartons, tins or similar containers –
  (a) are secured by laying up articles in a header and stretcher fashion and that corners are securely bonded; and
  (b) are stepped back half the depth of a single container at least every fifth tier or that, alternatively, successive tiers are stepped back by a lesser amount: Provided that at least the same average angle of inclination to the vertical is achieved; Provided further that where the containers are of a regular shape and their nature and size are such that the stack will be stable, they may be

stacked with the sides of the stack vertical if the total height of the stack does not exceed three times the smaller dimension of the underlying base of the stack.

(5) Notwithstanding the provisions of sub-regulation (4), freestanding stacks that are built with the aid of machinery may, with the approval of an inspector, be built to a height and in a manner permitted by the nature of the containers being stacked: Provided that –

(a) the stacks are stable and do not overhang: and

(b) the operator of the stacking machinery is rendered safe as regards falling articles.

7.5.2 SAFE STACKING GUIDELINES AND PROCEDURES

When stacking material on a construction site it can be dangerous if workers do not follow the H&S guidelines and safe working or operational procedures. One of the most common accidents related to storage is that of a stack collapsing. When construction materials are stacked too high or in an unstable arrangement, removing an item from the stack or bumping the stack can cause the rest of the materials to fall. If heavy objects are involved, this can pose a real threat to workers. Falling materials and collapsing loads can crush or pin workers, causing injuries or death.

Using an appropriate stacking method is one of the best ways to keep a stack from collapsing. The following are the most common options:

- **Block stacking** – Stack square items in a cube, making sure to secure them with strapping like wire or plastic shrink wrap.
- **Brick stacking** – To ensure even more security, turn each level of a stack 90 degrees. This helps hold the items in place should the stack be bumped.
- **Pinwheel stacking** – For even more protection than the brick pattern, turn each quadrant – not just each level – of items 90 degrees. Patterns like this help 'lock' everything in place.
- **Irregular stacking** – When dealing with irregularly shaped items, try adding sheets of plywood between each layer for added stability.

| Block pattern | Brick | Pinwheel | Irregular |

Figure 7.27: Various stacking methods

In addition to using secure stacking methods, the height and weight of the material being stacked needs to be considered to prevent a collapse. Heavy materials should generally be stacked close to the ground and not too high. Bricks, for example, should only be stacked 2 100 mm high, and if the pile is higher than 1 200 mm, the top of the pile should be tapered (50 mm back for every 300 mm of height above 1 200 mm).

To help prevent injuries when stacking construction materials on site, workers must do the following:
- Remove all nails from timber before stacking and do not stack more than 4 800 mm high (if workers will be handling it manually).
- Stack and level timber on solidly supported bracing.
- Ensure that stacks are stable and self-supporting.
- Use tapered stacking (a bit like a pyramid, with the layers getting narrower the higher up they are) for bagged materials; place bags in interlocking rows and cross-key them at least every ten layers.
- Stack bags and bundles in interlocking rows to keep them secure.
- Remove bags from the stack by starting from the top row first.
- Stack and block poles as well as structural steel, bar stock and other cylindrical materials to prevent spreading or tilting unless they are in racks.
- Observe height limitations when stacking materials.
- Consider the need for availability of material when storing and stacking them.

7.6 ELECTRICAL SAFETY

Electrocutions in construction represent a major challenge as many workers succumb to the effects of electrocution on construction sites. An electrical hazard occurs when a worker makes contact with an electrical conductor carrying electrical current and is at the same time in contact with the ground or another object that includes a conductive path to the ground. The body of the worker completes the electrical circuit enabling the electrical current to pass through him, usually with disastrous results.

Causes of electrical shock on construction sites include:
- Contact with bare, energised electric wires
- Electrical equipment that has not been properly grounded resulting in short circuits
- Working with electrical equipment on damp floors or wet surfaces
- Static electricity discharges
- Making use of metal ladders when working on electrical equipment resulting in a direct link from the electric power source to the ground
- Working on electrical equipment without shutting off the power supply
- Lightning strikes while working in the open.

Figure 7.28: Example of warning signage on electrical equipment

It is therefore critically important that all electrical conductors and equipment used by all contractors on the site must satisfy the approved specifications, standards and codes of the ECA (Electrical Contractors Association). All equipment must be approved and free of visible and identifiable hazards that might cause injuries or deaths. The two principal ways to prevent an electrical shock or electrocution are insulation and grounding. Insulation can become worn and deteriorated. Therefore, all equipment must be properly inspected before use. Prevention of shocks due to a faulty grounding system can be achieved through the use of a Ground-Fault Circuit Interrupter (GFCI) device, which is able to sense small levels of electrical current leakage to ground. The GFCI acts like a circuit breaker and shuts down the electricity supply.

7.6.1 LOCKOUT SYSTEMS

Lockout-tagout (LOTO) or lock and tag is a common safety procedure that ensures that dangerous machines and equipment are properly shut off and de-energised and not started up again prior to the completion of maintenance or servicing work. LOTO requires that electrical power sources be 'isolated and rendered inoperative' before any repair or maintenance procedure is started. 'Lock and tag' works in conjunction with a lock usually locking the device or the electrical power source with the hasp, and placing it in such a position that no electrical power sources feeding the device being worked on can be turned on. The procedure requires that a tag be affixed to the locked device indicating that it should not be turned on.

The lockout-tagout (LOTO) or lock and tag procedure is necessary to:
- minimise/manage risk of exposure to live energy sources and hazardous substances
- prevent accidental/unlawful machine or equipment start-up during maintenance or any other time
- comply with relevant legislation, regulations, standards and codes of practice.

THE SIX-STEP LOCKOUT/TAGOUT PROCEDURE

A lockout/tagout procedure or method should include the following six steps.

Step 1: Preparation – lockout/tagout

The first step in the locking and tagging out of equipment or a machine for service and maintenance is to prepare. During the preparation phase, the duly approved or authorised person, such as all authorised operator/drivers, mechanics/service crew and persons who inspect equipment, must investigate and obtain a complete understanding of all kinds of hazardous energy that must be controlled. The hazards must be identified to be able to control or monitor that energy.

Step 2: Shut down – lockout/tagout

When the planning has been completed, the actual process of powering down and locking out machines starts. At this point, the machine or equipment to be worked on must be shut down. All construction workers who will be affected by the shutdown must be notified.

Step 3: Isolation – lockout/tagout

The next step is to isolate the machine and or the equipment from any source of energy. This may involve turning off the electrical power at a breaker, or at the main isolator on the machine or equipment, or shutting a valve.

Step 4: Lockout/tagout

With the machine or equipment separated or isolated from its energy source, the next step is to lock and tag out the machine or equipment. During this step, the approved or authorised person will attach the lockout and tagout devices to the machine or equipment. This tag carries the name of the person who presented the lockout and additional information. The steps involved are as follows:
- Attach the scissors to the lockout point.
- Place the lock into the scissors and lock it.
- Keep the keys in a safe place.
- Team lock out.
- Use one scissors and multiple lock on one machine.

When two or more subcontractors are working on different parts of a larger overall system, the locked-out device is first secured with a folding scissors clamp that has many padlock holes capable of holding it closed. Each subcontractor applies their own padlock to the clamp. The locked-out device cannot be activated until all workers have signed off on their portion of the project and removed their padlock from the clamp. An example of lockout is shown in the figure.

Figure 7.29: Example of lockout

(Source: Anonymous)

Step 5: Stored energy check – lockout/tagout

During this step it is necessary to check for any hazardous energy that might still be 'stored' within the machine, or any 'residual' energy. In this case, any potentially hazardous stored or residual energy must be made non-hazardous in an appropriate manner using a safe working procedure.

Step 6: Isolation verification – lockout/tagout

In the final step the approved or authorised person verifies that the machine or equipment has been suitably isolated and de-energised.

7.6.2 THE EN 50110-1 SECURITY RULES

According to the European standard EN 50110-1, the security procedure before working on electric equipment comprises the following five steps:

1. Disconnect completely.
2. Secure against re-connection.
3. Verify that the installation is dead.
4. Carry out earthing and short-circuiting.
5. Provide protection against adjacent live parts.

7.7   FIRE PROTECTION AND PREVENTION

For a fire to start three elements are required, namely oxygen, fuel and heat. Since oxygen is naturally present in the atmosphere and environment, fire hazards generally involve the mishandling of fuel or heat. A fire is a chemical reaction between oxygen and a combustible fuel. Combustion occurs when the fire converts oxygen and the fuel into energy usually in the form of heat. For combustion to continue requires a constant source of oxygen, fuel and heat. Fires are classified according to their properties which are related to the nature of the fuel. Without a source of fuel there is no fire hazard. Fuels take many forms such as solids, liquids, vapours and gases. To avoid the negative outcomes of a fire, construction companies should establish a comprehensive fire safety programme, which should include assessment, planning, awareness/prevention and response. A fire committee should have members from all divisions of the company.

- Assessment involves continual assessment of construction sites for fire hazards by persons who are familiar with the fundamentals of fire hazard assessment.
- Planning involves the construction company having a fire safety plan that includes emergency escape procedures and routes, critical shutdown procedures, worker headcount procedures, rescue and medical procedures, procedures for reporting fires and emergencies and details of important contact personnel for more information when needed.
- Awareness/protection involves all workers in the construction company receiving awareness training so that they understand their role in executing the emergency plan. The awareness programme must be evaluated at regular intervals to ensure effectiveness and currency of the information and procedures.
- Response involves regular drills to ensure an effective and efficient response especially because panic sets in easily in an emergency. Emergency procedures must form a part of the induction programme.

Adequate portable fire extinguishers that are properly maintained and regularly tested must be positioned in strategic and well-marked locations on the construction site. Similarly, firehose reels must be placed in appropriate locations on the construction site. Proper visible signage must be placed strategically to direct workers to the location of the fire protection equipment. It is important to also provide contact information of emergency services placed in areas where they are visible and at least near a means of communication such as an alarm and telephone.

Care must be taken to properly store and handle flammable and combustible liquids on the construction site. The available supply or stock levels of these must be kept to a minimum. If a material is flammable, its labelling must be checked. Have any storage instructions been provided? Combustible materials cannot be stored near open flames, devices that might spark or areas where smoking is allowed. They should be stored in a flame-resistant cabinet isolated from locations on the site where construction activities take place. Flammable liquids must be sufficiently separated from other materials by a firewall. Some materials may interact with others in dangerous ways. When this is the case, these materials must be stored at a safe distance from each other. Prohibit smoking near any possible fuels. Consider a no smoking policy. Clean up all spills immediately after they occur. Always practise good housekeeping across the construction site. Repair all defective electrically powered tools, machines and equipment immediately. Use lockout and tagout procedures.

7.7.1 HOT WORK

In construction working with ignition sources near flammable materials is referred to as 'hot work.' Welding and cutting are examples of hot work. Fires are often the result of the 'quick five minute' job in areas not intended for welding or cutting. Getting a hot work permit before performing hot work is just one of the steps involved in a hot work management programme that helps to reduce the risk of starting a fire by welding or cutting in areas where there are flammable or combustible materials.

DEFINITION OF A HOT WORK MANAGEMENT PROGRAMME

Hot work management programmes are put in place to control or eliminate hot work hazards and their risks on construction sites. Programmes include the development of policies, procedures, and the assignment of responsibilities and accountabilities for all aspects of hot work. A programme includes:

1. Policies
   a. Where hot work is permitted
   b. When hot work is permitted
   c. Who authorises hot work.
2. Procedures
   a. What must be assessed before permitting/performing hot work in an area or on a process, piece of equipment or area
   b. What to do to prepare an area for hot work
   c. What to do if hot work cannot be avoided in a particularly hazardous area
   d. What hot work tools are required
   e. How to obtain a hot work permit, when they are required, and who can administer them.

3. Training
   a. Workers, supervisors, maintenance individuals, fire wardens, trained fire watch individuals, and contractors all have different roles, and must be trained accordingly.
4. Communications
   a. Posting procedures
   b. Posting policies
   c. Posting signs in areas that are prohibited from having hot work performed in them.

ALTERNATIVES TO HOT WORK

Hot work may be substituted with other methods, for example:

Instead of:	Use:
Saw or torch cutting	Manual hydraulic shears
Welding	Mechanical bolting
Sweat soldering	Screwed or flanged pipe
Torch of radial saw cutting	Mechanical pipe cutter

GENERAL GOOD PRACTICES BEFORE PERFORMING HOT WORK

Ensure that proper hot work procedures are strictly adhered to. The following should be considered:

- Ensure that all equipment is in good operating order before work on the construction site starts.
- Inspect the work area thoroughly before starting by checking for combustible materials in existing or completed structures on the site such as partitions, walls and ceilings.
- Remove any combustible materials on floors around the work zone. Combustible floors must be kept wet with water or covered with fire resistant blankets or damp sand.
- Use water ONLY if electrical circuits have been de-energised to prevent electrical shock.
- Remove any spilled grease, oil or other combustible liquid.
- Move all flammable and combustible materials a reasonable distance away from the work area.
- If combustibles cannot be moved, they should be covered with fire resistant blankets or shields.
- Block off cracks between flooring, along skirtings and walls, and under door openings, with a fire resistant material. Close doors and windows.
- Cover any existing wall or ceiling surfaces with a fire resistant and heat insulating material to prevent ignition and accumulation of heat.

- Secure, isolate and vent pressurised vessels, piping and equipment as needed before beginning hot work.
- Inspect the area following work to ensure that wall surfaces, partition studs, wires or dirt have not heated up.
- Vacuum away combustible debris from inside ventilation or other service duct openings to prevent ignition. Seal any cracks in ducts. Prevent sparks from entering the duct work. Cover duct openings with a fire resistant barrier and inspect the ducts after work has been completed.
- Post a trained fire watcher within the work area during welding, including during breaks, and for at least 30 to 60 minutes after work has stopped. Depending on the work done, the area may need to be monitored for up to three hours after the end of the hot work.
- Eliminate explosive atmospheres such as vapours or combustible dust or do not allow hot work. Shut down any process that produces combustible atmospheres, and continuously monitor the area for accumulation of combustible gases before, during and after hot work.
- If possible, schedule hot work during shutdown periods or when no construction activities are being executed.
- Comply with the required legislation and standards applicable to the workplace.

Figure 7.30 shows an example of a pro forma hot work permit.

HOT WORK PERMIT

1. This permit is valid for one job only.
2. This copy is to be retained by the person authorised to perform the hot work and must be produced on the request of any of our employees.
3. Management will retain a copy.
4. Description of work:

5 Exact location of the job:

6. Permit valid from   to
7. Note particular hazards or include hazardous areas list.

---

8. STANDARD PRECAUTIONS TO BE TAKEN		
Precaution	**Management requirement (✔)**	**Contractor to confirm (✔)**
Equipment in good working order		
All combustibles removed a minimum of 5 metres from the work area		
Fire extinguishers available		
Hosereel available		
Fire blanket provided		
Ventilation provided		
Barricades required		
Signs required		
Check the area for explosive atmosphere		
Post welding/cutting fire watch		
Additional precautions	**Management requirement (✔)**	**Contractor to confirm (✔)**

I verify that the job area has been examined and authorise hot work to be carried out, provided the above conditions are maintained throughout the term of the permit.

Name: Signed:
Name of person(s) doing the work:
Company name:
Signed:
Date:

On completion of the hot, work sign and return to place of registration.

I verify that the hot work has been completed in accordance with the authorised conditions outlined in the first part of this form.

Person doing hot work:

Signed:

Location of work inspected 30 minutes after job completed

Signed:

Figure 7.30: Example of a pro forma hot work permit

A written hot work permit needs to be issued by a supervisor who should examine the location on the construction site and methods of protection, and specify necessary precautions. The permit is issued for a specific time and area. Work should not begin in a new area without obtaining a new permit. A permit should not be issued unless the following requirements are met:

- Adequate fire extinguishing equipment is located at the work area on the construction site.
- Combustible materials have been relocated at least 11 m from the area of operation. If they cannot be relocated, they should be shielded by non-combustible welding curtains or fire-resistant blankets.
- Combustible floors and walls should be wet down with water if possible. A fire watch is trained and present at the site.
- The fire watch should be maintained for at least 30 minutes after operations have been completed and the area should be checked again at least one hour afterward.
- Work should only begin after ensuring that all equipment and surrounding conditions are safe.
- Continue to work only if approved work conditions remain unchanged.

The job is not considered complete until the responsible supervisor signs the hot work permit and checks the area. Many losses occur when the hot work permit is signed off without a final review of the area.

SPECIAL CONSIDERATIONS WHEN DOING HOT WORK
- Never cut or weld around explosive atmospheres or where combustible dusts or liquids are present. Explosive material, such as flammable liquids, gases, vapours or combustible dust, may reside in uncleaned or improperly prepared drums, tanks or other containers. Sparks can pass through cracks, or holes in walls, floors, broken windows, or open doorways.
- Before initiating hot work at roof level or on scaffolding, be sure that the area below is secured, and any sparks or hot debris will not cause a fire.
- Hot work permits should not be issued if any portion of the plant's fire protection systems is impaired.

7.8   PERSONAL PROTECTIVE CLOTHING AND EQUIPMENT

Unfortunately, it is common in the construction industry to use personal protective equipment or PPE as the primary control measure for personal safety. Instead, it should be a measure of last resort after exposure to hazards on the construction site has been mitigated as far as reasonably possible. Personal protective equipment (PPE) includes all items of clothing and work accessories that are specifically designed to protect construction workers from exposure to hazards present on the construction site that could potentially cause serious injuries on the site itself and illnesses subsequently arising from working on that construction project. Injuries and illnesses may result from contact by the construction worker with chemical, radiological, physical, electrical, mechanical or other workplace hazards. Generally, PPE may include gloves, safety glasses, steel-toed boots and footwear, goggles or face shields, hearing protection in the form of earplugs or muffs, hard hats, respirators, or overalls, coveralls, vests and full body suits. Therefore, it is important that all PPE should be safely designed and constructed and be maintained in a clean and reliable fashion. If the PPE does not fit properly, it can make the difference between being safely covered or dangerously exposed.

Protective equipment should not replace engineering, administrative or procedural control measures used to protect construction workers. All alternative ways of protecting workers on construction sites must be fully investigated and implemented before using PPE. When engineering, work practice and administrative controls are not feasible or do not provide sufficient protection, contractors must provide personal protective equipment to their workers and ensure its proper use. Therefore, PPE should always be a measure of last resort and not first resort as is typically the case in the construction industry. Where PPE is a measure of first resort it creates in the mind of the construction worker a false sense of security. Consequently, the worker assumes that the hazard has been mitigated and that there is no threat from exposure to it.

Figure 7.31: Lack of any protection for the workers exposed to this harsh construction activity on a site in India

Therefore, it is imperative that the hierarchy of H&S hazard controls, as advocated by the National Institute for Occupational Safety and Health (NIOSH) in the USA, should be used to mitigate exposure to construction site hazards. This hierarchy, which is often depicted as a pyramid, contains five levels and is meant to be worked from the top of the pyramid downward.

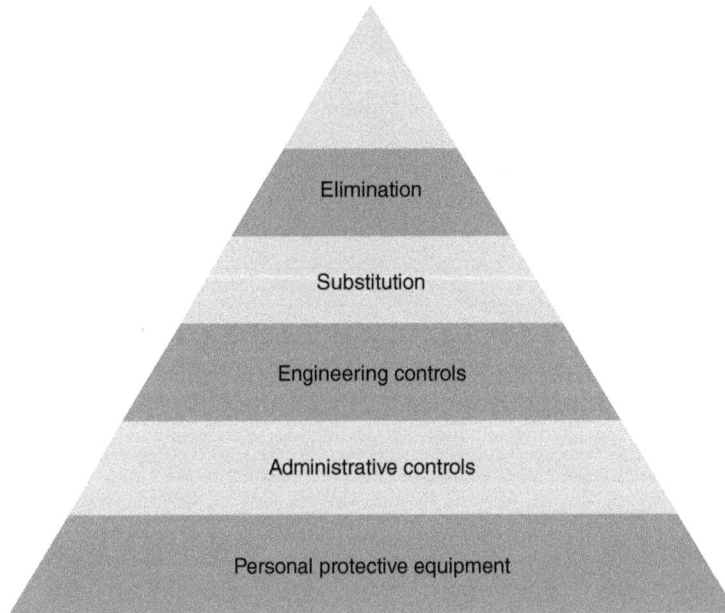

Figure 7.32: Hierarchy of H&S hazard controls

(Source: The National Institute for Occupational Safety and Health (NIOSH))

The five controls in the hierarchy are as follows:
1. Eliminating the hazard: The most effective way to keep construction workers safe from a hazard on the construction site is to eliminate the hazard altogether. If a tool or piece of equipment represents a danger to workers the necessity of the item should be considered and where not necessary, it should not be used at all.
2. Reducing the hazard (substitution): Where it is not viable to completely eliminate the hazard it should be substituted with something less dangerous. Even though the hazard still exists, workers will not be in nearly as much danger.
3. Implementing engineering controls: In cases where trying to eliminate or reduce the hazard on site is not viable, the next best option is to limit the exposure to the hazard by using engineering controls. Engineering controls are designed as modifications or designs to machines, equipment and processes that reduce the amount of exposure and physically protect the construction worker from harm.
4. Utilising administrative controls: Where it is not feasible to implement engineering controls, the next option in the hierarchy of hazards to be considered is to use administrative controls. These controls are at risk of human error and are used as a temporary intervention rather than a sustainable, long-term solution. These controls are typically rules and regulations that are put in place to minimise exposure to hazards. Examples of administrative controls include the following:

- Training: Workers should be trained to identify hazards, monitor hazard exposure, and implement health and safety procedures for working around the hazard. Additionally, workers should know how to protect themselves and their co-workers.
- Procedures: The steps in a job process may need to be rearranged or updated to keep the worker from encountering the hazard. Developing standardised safe work practices such as a safe work or operating procedure (SWP or SOP) is an important step.
- Maintenance: Having a properly designed, documented and implemented maintenance schedule for machines known to be hazardous is important in improving health and safety. Preventive maintenance will address any equipment issues before they become a problem.
- Housekeeping: Sustaining a clean and clutter-free space will greatly reduce the risk of injury and can minimise the severity of an accident.
- Signs: Wall and floor signs can be posted or installed to enforce administrative controls. Visual cues can remind workers which areas are prohibited from entering and, for example, when breaks need to be taken to limit heat exposure.

5. Personal protection equipment: Finally, PPE should be considered when all other options have been explored and exhausted. Although considered 'the last line of defence', or a measure of last resort, PPE can make the difference between a minor injury and a severe, life-threatening injury or even death.

Contractors are also required to train each construction worker required to use PPE to know:
- when PPE is necessary
- what kind of PPE is necessary
- how to properly put PPE on, adjust, wear and take it off
- the limitations of the PPE itself
- proper care, maintenance, useful life and disposal of the PPE

A risk assessment must be conducted by the contractor to identify the potential hazards that workers could be exposed to on the construction site and select the appropriate PPE for their adequate protection. Therefore, the construction workers should be trained with potential topics of knowledge, which comprise the proper PPE for their work, when this PPE should be used, how to wear, adjust, maintain and discard this equipment properly, and the shortcomings of the PPE.

It is critical therefore, to use or wear all the equipment, protective devices or clothing recommended on equipment, products or Material Safety Data Sheets (MSDSs) or as required by the contractor, the Act or any other relevant construction H&S regulation. Any broken, worn or missing PPE should be reported to the supervisor and where unsure about the most appropriate PPE for the construction activity, workers should request advice from the supervisor.

7.8.1 MAINTENANCE OF PERSONAL PROTECTIVE EQUIPMENT

All PPE should always be used and maintained in a sanitary and reliable condition. Maintenance must be carried out on, for example, respirators, safety shoes and gloves to ensure that the equipment will always function properly and increase their lifespan and use. Damaged or broken equipment must always be replaced. Slack, worn-out, sweat-soaked or twisted headbands of hardhats must be replaced regularly. A visual inspection of the hardhat can determine when the headband must be replaced.

7.8.2 CLEANING OF PERSONAL PROTECTIVE EQUIPMENT

PPE must always only be cleaned in accordance with the specifications of the manufacturer. Any other methods of cleaning will cause the equipment to malfunction and reduce its effectiveness.

7.8.3 STORAGE OF PERSONAL PROTECTIVE EQUIPMENT

Goggles should be kept in a case when not in use. Spectacles, in particular, should be given the same care as one's own glasses, since the frame, nose pads and temples can be damaged by rough usage. Items such as respirators and hearing protection equipment should be placed in a clean, dust-proof container, such as a box, bag or plastic envelope, to protect them until reissue. Store equipment only according to the specifications of the manufacturer.

Figure 7.33: Proper storage of PPE

(Source: https://nasdonline.org/7314/d002518/oregon-osha-pesticide-use-and-your-personal-protective.html)

PPE is designed to provide protection to a specific body part or several specific body parts. PPE only differs in the amount of protection it offers as well as the specific threat it protects against. It is therefore very important that the correct PPE is used to provide the protection required after implementing the hierarchy of control. Table 7.2, sourced and adapted from WorkSafe in New Zealand, lists the various body parts and the range of PPE that could be used to protect them.

Table 7.2: Body areas and range of PPE used for protection

Body area	Hazards	Potential clothing examples	Notes
Head and neck	• Impact from falling or flying objects, risk of head bumping, hair becoming entangled in machinery, chemical drips or splash, climate or temperature	• Face shields • Sunhats • Hard hats • Neck protection such as scarves for use during welding • Hairnets • Bump caps	• Some safety hard hats incorporate or can be fitted with specially designed eye or hearing protection
Eyes	• Chemical splash, dust, projectiles, gas and vapour	• Safety glasses and goggles • Face shields • Visors	• Make sure that the chosen eye protection has the right combination of impact/dust/splash protection for the task and fits the user properly
Ears	• Noise -a combination of sound level and duration of exposure, very high level sounds are hazardous even if for short duration	• Earplugs • Earmuffs • Semi-insert/canal caps	• Provide hearing protectors appropriate to the type of work and ensure that workers know how to fit, use and maintain them • Choose protectors that reduce noise to an acceptable level, while allowing for safety and communication

Body area	Hazards	Potential clothing examples	Notes
Hands and arms	• Abrasions, temperature extremes, cuts and punctures, impact, chemicals, electric shock, vibration	• Gloves made of latex, rubber, leather or metal • Gloves with a cuff • Long sleeved tops • Gauntlets and sleeving that covers part or all of the arm	• Chemicals quickly penetrate some materials – take care in selection • Barrier creams are unreliable and are no substitute for proper PPE • Wearing gloves for long periods can make the skin hot and sweaty, leading to skin problems. Using separate cotton inner gloves can help prevent this challenge
Feet and legs	• Slips, trips and falls, abrasions, temperature extremes, cuts and punctures, impact, chemicals, electric shock, vibration	• Rubber boots • Thermally insulated boots • Safety boots with protective toecaps and penetration-resistant mid-soles • Foundry boots • Chainsaw boots • Anti-static, electrically conductive boots • Boots with oil/chemical resistant soles • Non-slip shoes	• Wearing boots for long periods can make the skin hot and sweaty, leading to skin problems. • Ensure that the correct size footwear is worn to prevent them becoming a hazard in themselves
Whole body		• Conventional or disposable overalls • Aprons • Chemical suits • Cooling vests • Weather-proof gear – water-proof trousers, rain coats • Fire-proof clothing • Hi-visibility (hi-viz) clothing	

(Source: Adapted from https://worksafe.govt.nz/topic-and-industry/personal-protective-equipment-ppe/protective-clothing/)

REVIEW QUESTIONS

1. What are some of the steps that can be taken to prevent excavation-related incidents from occurring on the construction site?
2. What are basic health and safety rules when using ladders?
3. What are some of the precautions to be taken when working with cranes?
4. How can falls from scaffolds be prevented?
5. What are the design ratios for a tower scaffold?
6. What are some of the special precautions that need to be considered when making use of material and a passenger hoist on the construction sites?
7. Why is good housekeeping necessary?
8. How does a GFCI device provide protection against electrical shocks?
9. Outline the details of an effective fire safety programme.
10. Describe the possible stacking options for materials on a construction site.
11. Cite examples of good and poor housekeeping.
12. What must be considered when stacking timber or lumber?
13. How would one safely store chemical substances?
14. Who should be responsible for lockout and tagout procedures?
15. When would a hot work permit system be used?

REFERENCES

Balamurugan, P, Muthamilselvi, P & Balashanmugam, P. 2019. Effective work permit system to minimize the hazards in eid parry (India) limited. *International Journal of Engineering, Science and Mathematics*, 8(8): 20–42

British Compressed Gases Association, 2017. Code of Practice 7 The safe use of oxy-fuel gas equipment (individual portable or mobile cylinder supply). Available at: https://pgstraining.com/wp-content/uploads/2019/04/CP7-Rev-7-2014-Update-Jan-2017.pdf

BOC. nd. *Storage of Cylinders*. [online] BOC Zimbabwe Gases. Available at: http://www.boc-gas.co.zw/en/health_and_safety/gases_and_cylinders/storage_cylinders/index.html

Canadian Centre for Occupational Health and Safety. nd. Available at: https://www.ccohs.ca/oshanswers/chemicals/compressed/compress.html

Canadian Centre for Occupational Health and Safety. nd. Available at: https://www.ccohs.ca/oshanswers/chemicals/howto/gas_cylinder.html

Canadian Centre for Occupational Health and Safety. nd. How Do I Work Safely With - Compressed Gases. Available at: https://www.ccohs.ca/oshanswers/prevention/comp_gas.html

Canadian Centre for Occupational Health and Safety. nd. Available at: https://www.ccohs.ca/oshanswers/safety_haz/ladders/step.html

Canadian Centre for Occupational Health and Safety. nd. Available at: https://www.ccohs.ca/oshanswers/safety_haz/conveyor_safety.html

Chakraborty, A. 2019. Construction safety Management. *International Journal of Town Planning and Management*, 5(1): 12–20

Chang W, Leclercq S, Lockhart TE & Haslam R. 2016. State of science: Occupational slips, trips and falls on the same level. *Ergonomics*, 59(7): 861–883. DOI: 10.1080/00140139.2016.1157214

Department of Labour. 1993. Occupational Health and Safety Act 85 of 1993.

Goetsch, DL. 2011. *Construction Safety and Health*. Upper Saddle River, New Jersey: Pearson Education Inc

Haupt, TC. 2021. *Management of Safety, Health and Environment in South Africa: A Handbook*. Newcastle upon Tyne: Cambridge Scholar Publishing

Health and Safety Executive. 2012. Safety in gas welding, cutting and similar processes. Available at: https://www.hse.gov.uk/pubns/indg297.pdf

Holt, Allan St John. 2001. *Principles of Construction Safety*. Osney Mead, Oxford: Blackwell Science Ltd

Health and Safety Authority (HSA). 2008. [online] Australianscaffolds.com.au. Available at: http://australianscaffolds.com.au/documents/Scaffolding_COP.pdf

Health and Safety Authority (HSA). nd. [online] Besmart.ie. Available at: https://www.besmart.ie/fs/doc/Construction/Construction_Documents/scaffolds_PoP_v11.pdf

Health and Safety Executive (HSE). nd. HSE safety alert on the use of tower cranes. https://www.mantiscranes.ie/wp-content/uploads/2017/01/CPA-TCIG-HSE-Safety-Alert-Use-of-Tower-Cranes.pdf

HSE Books. 2002. Ergonomics evaluation into the safety of stepladders. Literature and standards review Phase 1 CRR418

HSE Books. 2002. Ergonomics evaluation into the safety of stepladders. User profile and dynamic testing - Phase 2 CRR423

HSE Books. 2004. Evaluating the performance and effectiveness of ladder stability devices RR205

HSE Books. 2005. A toolbox talk on leaning ladder and stepladder safety Leaflet INDG403

HSE Books. 2005. Top tips for ladder safety Pocket card INDG405

Kukfisz, B, Ptak, S, Półka, M & Woliński, M. 2018. Fire and explosion hazards caused by oxygen cylinders. *WIT Transactions on the Built Environment*, 174: 141–151

Kumar, P. 2017. Significance of fire safety in pictorial manner. *Fire Engineer*, 42(3): 13–16

Lingard, H, Cooke, T, Zelic, G & Harley, J. 2021. A qualitative analysis of crane safety incident causation in the Australian construction industry. *Safety Science*, 133

Mahto, DG. 2016. Productivity safety and test of hypothesis: An analysis of significance of good housekeeping and safety initiatives in industries. Available at SSRN 2773163

McKinnon, RC. 2019. *The Design, Implementation, and Audit of Occupational Health and Safety Management Systems*. CRC Press

Mining Safety. nd. *Mining Safety | Lock Out Machinery in the Mining Industry*. [online] Miningsafety.co.za. Available at: https://www.miningsafety.co.za/dynamiccontent/84/dynamiccontent/69/Physical-Fitness

Nadeau, S, Kenné, J, Emami-Mehrgani, B & Badri, A. 2016. Advances in integration of equipment lockout/tagout, determination of actual production capacity and production/maintenance planning. *Safety Science Monitor*, 19(2)

Nawi, MNM, Ibrahim, SH, Affandi, R, Rosli, A & Basri, FM. 2016. Factor affecting safety performance construction industry. *International Review of Management and Marketing*, 6(8S): 280–285

Occupational Safety and Health Administration. 2002. Materials handling and storing. Washington, DC: U.S. Dept. of Labor, Occupational Safety and Health Administration

Occupational Safety and Health Administration. nd. *Portable Ladder Safety*. [online] Osha.gov. Available at: https://www.osha.gov/Publications/portable_ladder_qc.html; and https://www.cpa.uk.net/tower-crane-interest-group-tcig-publications/

Prince, PJ. 2017, January. Machine risk assessment and reduction, an ongoing journey. In ASSE Professional Development Conference and Exposition. American Society of Safety Engineers

Rinawati, S. 2018. Level of safe behavior with the implementation of hot work permit approach in PT Bbb East Java. *Journal of Vocational Health Studies*, 1(3): 89–96

Safe at Work California. nd. Conveyors - a Safety Message. https://www.safeatworkca.com/safety-articles/conveying-a-safety-message/

Safe Stacking and Storage. 1999. [online] Cdn.auckland.ac.nz. Available at: https://cdn.auckland.ac.nz/assets/science/for/current-students/HR/health-safety-wellness/documents/SafestackingandStorage.pdf

Safety & Health Procedures Manual. 2004. [online] Co.ontario.ny.us. Available at: http://co.ontario.ny.us/DocumentCenter/View/2292/Safety--Health-Manual?bidId=

Sowa, M. 2018. Threats to health and safety at work when completing a warehouse process. *European Journal of Service Management*, 27: 285–291

Spellman, FR & Whiting, NE. 1999. *Machine Guarding Handbook: A Practical Guide to OSHA Compliance and Injury Prevention*. Government Institutes

Stacking and Storage of Materials. [online] qushum.co.za/Safety/Docs/C1/V3QushumChecklist … Available at: https://www.safetyblognews.com/safe-stacking-and-storage-in-the-warehouse/

Stewart, RA. 2020. *Understanding how Safety Posters Affect Perception of Safety Culture using Virtual Reality*. Theses and Dissertations, Mississippi State University. Available at: https://scholarsjunction.msstate.edu/td/4825

Virginia Commonwealth University. nd. [online] Safety.louisiana.edu. Available at: https://safety.louisiana.edu/sites/safety/files/Compressedgascylindersafety.pdf

Yamin, SC, Parker, DL, Xi, M & Stanley, R. 2017. Self-audit of lockout/tagout in manufacturing workplaces: A pilot study. *American Journal of Industrial Medicine*, 60(5): 504–509

Yan, CK, Siong, PH, Kidam, K, Ali MW, Hassim, M, Kamaruddin, MJ & Kamarden, H. 2017. Contribution of permit to work to process safety accident in the chemical process industry. *Chemical Engineering Transactions*, 56: 883–888

CHAPTER 8
MANAGING COMMON
CONSTRUCTION HEALTH HAZARDS

8.1 INTRODUCTION

The construction industry is notorious for exposing those who work in it to harsh working conditions with little regard for the consequences of these exposures to their health and overall well-being. This lack of consideration for the effects of construction work activities on the health of the workforce can be attributed to the difficulty in linking the onset of an illness or disease to exposure to conditions on construction sites and to a particular contractor for whom the worker may have worked. In most cases, symptoms are not generally visible. Compounding this challenge is the time lag between exposure when a worker may be pre-symptomatic and the manifestation of the symptoms of work-related illness or disease. Therefore, contractors are reluctant to accept responsibility for what they do not see as directly related to them. Instead, they focus on what is more easily managed, namely the safety of the workers; they neglect the health aspects, giving credence to the claim that 'health is the second cousin to safety'. However, contractors are increasingly being held responsible for managing exposures on their construction projects that might eventually lead to health problems and conditions in their workforce. The COVID-19 pandemic has enforced a shift by making contractors more aware of their responsibility to fulfil their constitutional obligations to their workers. All construction employers have not only a moral obligation but a legal obligation in terms of the national Constitution to provide workplaces and work environments that are safe and do not present any threat to the health and well-being of their workers. Some contractors take these obligations more seriously than others.

It is important to be aware that a construction project comprises the following four basic elements, all of which must be addressed by the contractor:
1. The construction worker
2. The tools or equipment used to perform construction activities
3. The construction work activity
4. The work environment on the construction site in which construction work activities are executed.

Occupational health has been defined as the multidisciplinary approach to the recognition, diagnosis, treatment, and prevention and control of work-related diseases, injuries and other conditions. Occupational illnesses and diseases tend to be chronic and generally characterised by temporary or permanent physical dysfunction which will affect the productivity and overall sense of well-being of construction workers.

According to the World Health Organisation (WHO), occupational health is made up of following three important legs:

1. Occupational Hygiene
2. Occupational Medicine
3. Occupational Safety

On the other hand, in South Africa occupational health is seen slightly differently and includes:

- Occupational Hygiene
- Occupational Medicine
- Primary Health Care

Occupational medicine and primary health care are practised by the occupational health medical practitioner and the occupational health nursing practitioner, while occupational hygiene is a specialised field practised by qualified occupational hygienists. A total health care programme in a construction organisation can be developed and integrated when all these professionals in their various roles and functions work together to develop a holistic construction occupational health strategy in an environment where occupational health is not just a priority but a value of the organisation. Priorities change. Values tend to remain constant.

8.2 RELATIONSHIP BETWEEN WORK AND HEALTH

The following diagram illustrates that construction work and the construction working environment on sites will have an impact on construction worker health and well-being. The converse is also true, namely that the health of construction workers will have an impact on the work they do and eventually on the overall well-being of the contractor. Because of this relationship it is important for construction organisations not only to take an interest in the health of their workers, which is required in any case in terms of the Constitution, but also to ensure that their health is being taken care of as far as is reasonably possible.

Importantly, it is not necessarily the current or present occupation of a worker that is the cause of the occupational disease, so all previous occupations should be considered since the disease or affliction may have been contracted during a previous period of employment. For example, a person could have been exposed to asbestos at a previous place of work, and is not exposed to it currently, but only now presents with cancer of the lungs in the present workplace, since the effects of the previous exposure may be latent for up to 50 years.

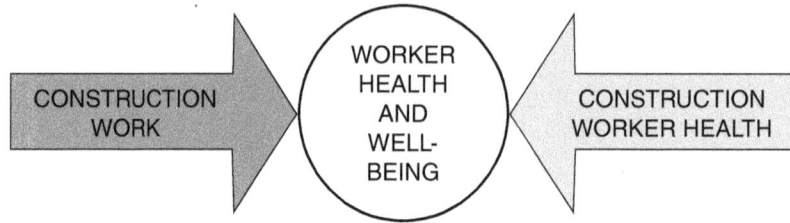

Figure 8.1: Construction work and the construction working environment have an impact on construction worker health and well-being

Apart from the previous occupations of workers, their hobbies or recreational activities could also be responsible for the contracting of an occupational health disease or risk. Workers who, for instance, work in their garden at home may be exposed to insecticides and pesticides without adequate protection. The effects of health at work, in converse, can typically be considered at three levels:

1. Workers who have no apparent health problems
2. Workers whose work ability have been impaired by illness or injury
3. Workers whose health may impact the health and safety of fellow workers or their community.

An example of the last level could be a worker who was exposed to the coronavirus coming to work and passing the virus to fellow workers through contact resulting in the spread of the infection to their respective families and communities.

8.3 OCCUPATIONAL HEALTH CHALLENGES

According to the World Health Organization, the most important occupational health challenges for the future include the following:

- Occupational health problems linked to new information technologies and automation
- New chemical substances and physical energies
- Health hazards associated with new biotechnologies
- Transfer of hazardous technologies
- Ageing working populations
- Special problems of vulnerable and underserved groups such as those who are chronically ill, handicapped and disabled, including migrants and the unemployed
- Problems related to growing mobility of worker populations and occurrence of new occupational diseases of various origins such as various coronaviruses in recent years.

8.4 OCCUPATIONAL HYGIENE STRESSORS

The environmental stressors of occupational hygiene or causes of the health problems of construction workers may be divided into the following five categories:
1. Biological
2. Chemical
3. Physical
4. Ergonomic
5. Psychosocial.

Occupational hygiene uses science and engineering to prevent ill health caused by the working environment. It helps both contractors and their workers to understand the risks, thus enabling them to improve working conditions and working practices.

Work has always involved hazards to health, for example:
• Plumbers have been poisoned by the lead they used for pipes and joints when lead was the material of choice.
• Operators of plate compactors and jack hammers experience lengthy exposure to excessive vibration.
• Roofers are exposed to friable asbestos fibres and dust, resulting in mesothelioma and lung cancer.

With good occupational hygiene practice, some historical risks have been eliminated and others brought under control. COVID-19 has elevated the need for and importance of both personal and occupational hygiene to prevent the spread and rate of infection in the workplace. Consequently, employers have additional responsibilities imposed on them.

8.5 THE OCCUPATIONAL HEALTH MODEL

The occupational health model shown in Figure 8.2 highlights the various functions that organisational health plays in South Africa.

This particular model integrates the five key areas of health delivery in the workplace and was designed and implemented with great success by the Chamber of Mines in the late 1980s. Many industrial and commercial organisations have adapted the model to meet their own health-related needs.
1. Primary health care includes all preventive and promotive health, health interventions, treatment of minor ailments and trauma. Preventive and promotive health includes health education and screening, while health interventions would include things like family planning and immunisation.
2. Occupational health is a specialist branch of medicine that focuses on the physical and mental well-being of workers in the workplace.
3. The Employee Assistance Programme (EAP) leg of the occupational health model in South Africa is a psycho-social support programme for workers.

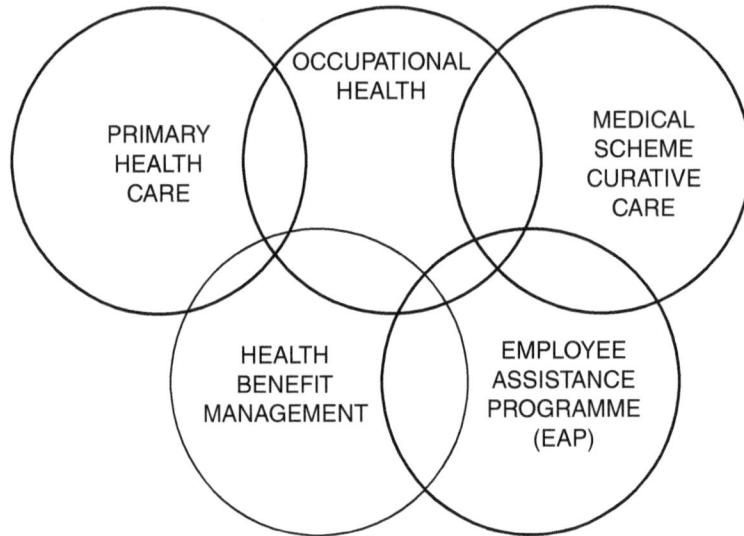

Figure 8.2: Five key areas of health delivery in the workplace

4. The health benefit management leg of the occupational health model in South Africa includes all different types of input and advice given to workers and management with regards to absenteeism, disability and incapacity management, COIDA interaction with medical aid and medical adjudication.

5. The medical scheme and curative care leg of the occupational health model in South Africa includes all the clinical services given to manage HIV, COVID-19, cannabis and other substance abuse, including counselling, support and wellness management.

The integrated delivery of service allows organisations to meet the challenges of complex health issues such as HIV and AIDS, COVID-19, silicosis, asbestosis, cannabis use, incapacity management and the management of human capital and resources.

8.6 MUSCULOSKELETAL AND SENSORY EFFECTS

Many construction activities involve repetitious body movements. Executing these activities and working continuously in static positions especially on elevated scaffold platforms, for example, kneeling for extended periods of time are major contributors to repetitive movement disorders that include meniscus lesions, bursitis, tendonitis, ganglion cyst, trigger finger and osteoarthritis of the knees. Other contributory movements include repetitive and heavy lifting, bending and twisting, frequent repetition of a movement, uncomfortable working position, exerting more force than necessary, and working for long periods without any breaks. Carpal tunnel syndrome (CTS) manifests as tingling, numbness, weakness or pain in the fingers or hand due to pressure on the median nerve in the wrist. Contributory causes include repetitively making the same hand and/or wrist movements as well as wrist injuries and spurs. These are common in the case of carpenters and bricklayers.

Lower back pain, especially in the case of bricklayers, is caused by some of the following:
- Lumbar strain where ligaments, tendons and/or muscles of the lower back are stretched
- Nerve irritation of the lumbar spine
- Lumbar radiculopathy caused by damage to the discs between the vertebrae
- Bone encroachment where the space for the adjacent spinal cord and nerves is reduced
- Bone and joint conditions that arise from injury such as fractures.

Slipped or herniated discs, where the tissue that separates the vertebral bones of the spinal column has been ruptured, manifest as a result of repetitive body movements during construction activities. Workers such as bricklayers, floor tilers, plasterers, carpenters and painters experience pressure against one or more spinal nerves, causing a shooting pain known as sciatica, weakness or numbness in the area served by these nerves. Treatment is in the form of anti-inflammatory and muscle-relaxant medications, exercises and physical therapy. Surgery may unfortunately be required in severe cases.

8.7 MIGRANT WORKERS, XENOPHOBIA AND VIOLENCE

Since South Africa opened its economy post-1994 the country has become an attractive destination for migrants especially from the 16 countries that are members of the South African Development Community (SADC). Both internal migrants within the borders of the country and international migrants face daily stresses associated with the challenges of moving to a new area or region, seeking work, struggling to access safe housing and a secure livelihood, feeling alone and without social support. Maternal, new-born and child health; HIV, COVID-19 infections and TB; non-communicable diseases; and violence and injury all affect migrants who enter workplaces in South Africa. It is hard to track workers who migrate as part of survival. Follow up medical appointments are easily missed and prescribed medication courses are not completed. As a consequence, migrants bring adverse health conditions into the workplace. Some diseases require lengthy periods of treatment, which are interrupted when workers migrate to another location and workplace in that location. Very often these workers do not have their medical history and records available for their new employer, thus creating a risk of having pre-existing health conditions introduced into the workplace. It is therefore necessary for the health system at work to recognise these challenges and develop a migration-aware health system.

Recently, xenophobia and xenophobic attacks on foreign migrants have increased in many parts of South Africa. The construction industry has not been unaffected. Xenophobia is defined described as the fear or hatred of foreigners or strangers; it is embodied in discriminatory attitudes and behaviour, and often culminates in violence, abuses of all types, and exhibitions of hatred. Many of these attacks have been extremely violent, with damage to the property of the victims and injuries to the victims themselves. The consequences of xenophobia include discrimination and barriers to accessing education, healthcare, employment and housing.

Workplace violence irrespective of its causes and reasons has become another challenge to the construction industry. The prevention of violence on construction sites involves a comprehensive approach that includes hazard analysis, record analysis and tracking, trend monitoring, incident analysis, prevention strategies, emergency response and worker training. The legal implications of violence in the workplace must be considered, which revolve around the rights of the violent perpetrators as well as their fellow workers. It is possible that these conflicting legal rights could create liabilities for contractors that are affected. It is critical to deal consistently and promptly with all incidences of violence.

The following guidelines to deal with violence generally and xenophobic attacks specifically may be helpful:
- Respond immediately to imminent threats or dangerous situations which could easily escalate and get out of hand.
- Take all threats seriously and investigate them because seemingly frivolous allegations might prove to be real.
- Encourage workers to report suspicious workers or activities or potentially threatening situations.
- Take disciplinary action when it is necessary after consultation with various worker representative forums.
- Establish policies that include ground rules for worker behaviour and responses in threatening and violent situations.
- Control access to all working areas on the construction site.
- Adopt a zero-tolerance policy towards threatening or violent behaviour.
- Provide support for both primary and secondary victims.
- Do everything necessary to restore the work environment on the construction site to normal following an incident.
- Use counselling to deal with aggressive workers that includes aggression management or even employment termination.
- Set up a crisis-management team charged with preventing and responding to violence on the site.
- Consider setting up an emergency response team to deal with the immediate trauma of what has happened.
- Demonstrate management commitment and involvement together with worker engagement and involvement.

8.8 ERGONOMICS

The Ergonomics Regulations were gazetted on 6 December 2019 by the Department of Employment and Labour under section 43 of the Occupational Health and Safety

Act 85 of 1993. Ergonomics, also known as human engineering or human factors engineering, is the multi-disciplinary science dealing with the interactions between people, and machines and their work environment. The Ergonomics Regulations provide guidelines for the management of ergonomics within organisations and also present an opportunity for organisations to benefit from the dual objective of ergonomics which is to improve worker health and safety while also optimising work output.

The official International Ergonomics Association (IEA) definition of ergonomics is as follows:

> Ergonomics is the scientific discipline concerned with the understanding of interactions among humans and other elements of a system, and the profession that applies theory, principles, data and methods to design in order to optimise human well-being and overall system performance (www.iea.cc).

Ergonomics seeks to improve the performance of systems by improving human–machine interaction. It involves the science of designing machines, products and systems to maximise the safety, comfort and efficiency of the people who use them. Ergonomists view workers and the objects they use in the workplace as one unit, and consequently ergonomic design blends the best abilities of people and machines. An ergonomically designed system provides optimum performance because it takes advantage of the strengths and weaknesses of both its human and machine components.

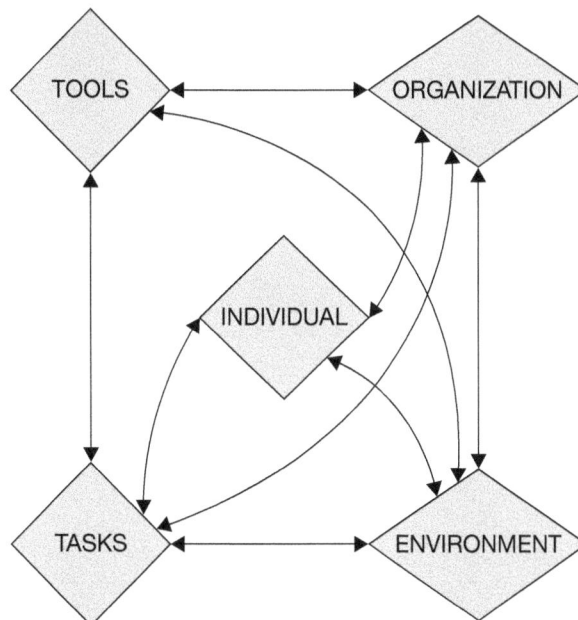

Figure 8.3: A model of an ergonomically designed system

Ergonomists draw on the principles of industrial engineering, psychology, anthropometry or the science of human measurement, and biomechanics, which is the study of muscular activity, to adapt the design of construction workplaces to the sizes and shapes of construction workers and their physical strengths and limitations. They also consider the speed with which construction workers react and how they process information, and their capacities for dealing with psychological factors, such as stress or isolation.

In general, ergonomics deals with the interaction between humans and additional environmental elements such as heat, light, sound, atmospheric contaminants and all tools and equipment pertaining to the construction site. The advantages of ergonomics or the proper designing of construction work systems and activities based on human factors include:

- more efficient operations
- fewer accidents
- reduced training time
- lower costs of production
- more effective use of construction workers.

Systems can be improved by:

- changing the task to make it more compatible with the user characteristics, for example, a redesigned scaffold for bricklayers
- changing the way work is organised to accommodate the psychological and social needs of workers by, for example, reducing the length of time that workers have to spend in confined spaces by introducing rotations and rest periods
- changing the work environment to make it healthier and safer and more appropriate for the task by, for example, improving the level of lighting in a darkish work area on the construction site.

8.8.1 ERGONOMIC HAZARDS

It is claimed that between 10% and 30% of the workforce in industrial countries and between 50% and 70% in developing countries such as South Africa may be exposed to heavy physical workload or to ergonomic working conditions such as lifting and moving heavy items or repetitive manual tasks. Repetitive tasks and static muscular load are found in construction on almost if not all construction sites. In many countries musculoskeletal disorders are the main cause of both short-term and permanent work disability.

Postural Stress is the term used to denote the mechanical load on the body by virtue of its posture. Postural stress can cause pain. People who work with the spine flexed forward about 60° for more than 5% of the day or 30° for more than 10% of the day are prone to suffer back pain.

A stable posture can only be kept if the different parts of the body are supported and maintained in an appropriate relation to their feet. For example: when a standing person leans forward to touch their toes, the pelvis moves backwards to compensate for the forward displacement of the centre of gravity of the upper body. In order to reduce postural stress, it is important to provide sufficient space around standing operators and plenty of room for the feet if losses of balance are to be avoided.

By doing the manual work in the wrong way, unnecessary stress and damage will be done to the soft tissue in the back of a construction worker such as a labourer or bricklayer, and can cause permanent back damage. It is extremely important to train construction workers who do hard manual work how to handle their bodies to avoid damage. Training should be done annually to refresh the memories of workers, and managers and supervisors should always be on the look-out for people doing harm to their bodies.

8.8.2 ERGONOMIC CONSIDERATIONS

Matching workers with their work

It is important to match construction worker capabilities with the related requirements of a given construction activity or task. If the job demands are equal to the capabilities of the worker or if they exceed them, the worker will be under too much strain and may not be able to perform the job, activity or task.

Work classification

Generally, the work demands are classified from light work to extremely heavy work in terms of energy expenditures per minute and the relative heart rate in beats per minute. The energy requirement for light work is 2.5 Kcal/minute and the heart rate is 90 beats per minute, while for extremely heavy work the energy requirement is 15 Kcal/minute or the equivalent of 6 times higher and the heart beat is 160/minute or the equivalent of 1.75 times faster.

Workstation design

The workstation is the immediate area where the construction worker performs his or her work activities. The goal of designing a workstation such as a special bricklaying scaffold is to promote ease and efficiency of the performance of the bricklaying crew. Productivity will be compromised if, for example, the worker is uncomfortable and the workstation is poorly designed.

Workplace design

Given that the workplace is where the worker performs his or her duties, the most basic requirement is that it must accommodate the construction worker working in it. Specifically, this means that:
1. The workspace for the hands should be between hip and chest height in front of the body.

2. Lower location is preferred for heavy manual work.
3. Higher locations are preferred for tasks that require close visual observations.

The workplace should be designed relating the physical characteristics and capabilities of the worker to the design of equipment and to the layout of the workplace. When this match is achieved, there is arguably:
- an increase in worker efficiency
- a decrease in human error
- consequent reduction in accident frequency.

Workspace dimension

The dimension of the workspace can be grouped into three basic categories, namely minimal, maximal and adjustable dimensions. Minimal workspace provides clearance for ingress and egress in walkways and doors. Maximal workspace dimensions permit smaller workers to see the equipment by selecting a workspace dimension over which a small person can reach or by establishing control forces that are small enough so that even a weak person can operate the equipment. Adjustable dimensions permit the construction worker to modify the work environment and equipment so that it conforms to those workers on a particular set of anthropometric characteristics.

8.9 CORONAVIRUS INFECTIONS

A novel coronavirus (COVID-19) that emerged at a market in Wu Han, China, in December 2019 has since focused global attention on human coronavirus infections. Sadly, this outbreak has caused unprecedented numbers of infections and deaths everywhere. COVID-19 is the seventh identified human coronavirus and appears to have notable similarities to two other highly pathogenic human respiratory coronaviruses, namely severe acute respiratory syndrome coronavirus (SARS-CoV) and Middle East respiratory syndrome coronavirus (MERS-CoV), both of which have generated large-scale public health responses in the last 20 years. The novel coronavirus (nCoV) which was previously named 2019-nCoV, now designated COVID-19 has not been previously identified in humans. This virus can be spread by human-to-human transmission via respiratory droplets or direct contact with an infected individual. Tremendous worldwide efforts are being made to understand the molecular and clinical virology of this virus. Unbelievable molecular knowledge about the genomics, structure, and virulence of this virus has already been achieved. Further, due to large-scale international collaboration, investment and effort, a number of vaccines have been developed within record time to protect against or limit the infection of people across the globe. However, the long-term effectiveness of these has not been established yet.

Common signs of COVID-19 infection include respiratory symptoms, fever, cough, shortness of breath and breathing difficulties. Severe cases result in infections such as pneumonia, severe acute respiratory syndrome, kidney failure and even death. In China, most patients had presented with fever (83%–98%) or cough (76%–82%) and roughly one-third had shortness of breath. Less common symptoms included myalgias (11%), rhinorrhoea (10%), headache (8%), chest pain (2%) and gastrointestinal symptoms Communicable Diseases Center (3%). The promotion of hand hygiene and respiratory hygiene are essential preventive measures. The following flowchart was adapted from the one recommended by the CDC.

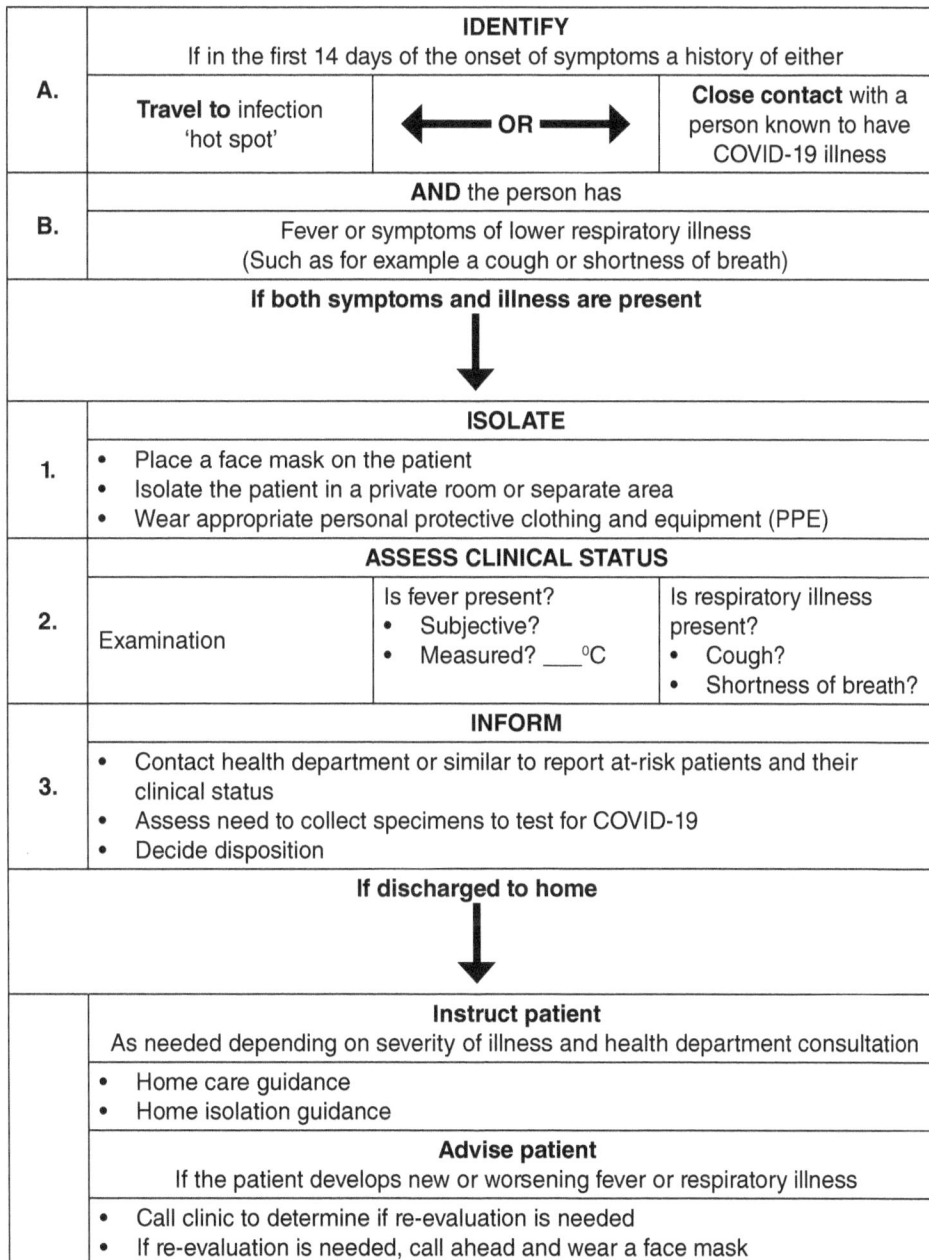

A.	**IDENTIFY** If in the first 14 days of the onset of symptoms a history of either		
	Travel to infection 'hot spot'	← **OR** →	**Close contact** with a person known to have COVID-19 illness
B.	**AND** the person has		
	Fever or symptoms of lower respiratory illness (Such as for example a cough or shortness of breath)		
	If both symptoms and illness are present ↓		
1.	**ISOLATE**		
	• Place a face mask on the patient • Isolate the patient in a private room or separate area • Wear appropriate personal protective clothing and equipment (PPE)		
2.	**ASSESS CLINICAL STATUS**		
	Examination	Is fever present? • Subjective? • Measured? ___°C	Is respiratory illness present? • Cough? • Shortness of breath?
3.	**INFORM**		
	• Contact health department or similar to report at-risk patients and their clinical status • Assess need to collect specimens to test for COVID-19 • Decide disposition		
	If discharged to home ↓		
	Instruct patient As needed depending on severity of illness and health department consultation		
	• Home care guidance • Home isolation guidance		
	Advise patient If the patient develops new or worsening fever or respiratory illness		
	• Call clinic to determine if re-evaluation is needed • If re-evaluation is needed, call ahead and wear a face mask		

Figure 8.4: Flowchart on advisable procedure around COVID-19

Given that every industry, including the construction sector, has been affected by COVID-19, it is important that construction organisations acquaint themselves with the most current legislation, regulations and protocols and proactively consider taking the following actions:

- Adapt and apply action steps through ongoing two-way communication and engagement between employer and workers that regularly address misunderstandings, misinformation, rumours and frequently asked questions.
- Encourage everyone to adopt protective behaviours.
- Provide information and guidance.
- Create a cooperative, harmonious and humane working environment with respect to open two-way communication with workers to understand and respond to their concerns, attitudes, beliefs and barriers to following health guidance and flexibility of working arrangements to accommodate threats of COVID-19 infections, self-isolation and quarantine without fear or panic.
- Liaise with employer–worker health and safety forums, unions, and other worker representative bodies to ensure informed collective decision-making about appropriate steps and actions to take to deal with COVID-19 effectively.
- Particular attention must be given to workers with reduced immune systems, including those who are HIV positive, diabetics, and workers with cancer, heart or lung complaints and other chronic diseases.
- Implement non-punitive 'emergency sick leave' policies.
- Implement flexible work hours, staggered shifts or working at home to enable additional workspace for social distancing.
- Avoid gatherings during lunch and tea breaks. Create a roster with flexible lunch times to avoid crowding.
- Increase welfare and hygiene facilities and ensure that they are properly stocked with soap and sanitisers and regularly deep cleaned.
- Provide PPE to all workers in the form of masks or transparent face shields, soap and water or 70% alcohol-based sanitiser and gloves as needed.
- Review all contractual arrangements to understand options now and in the future.
- Clarify the position over suspension of any contracts, orders and projects.
- Review H&S critical elements of projects and workplaces.
- Review production and delivery programmes and schedules.
- Identify if any activities can continue remotely or be brought forward.
- Ensure remote access to critical business and operational information is in place.
- Provide the means to operate from remote locations by providing workers who need them with internet and Wi-Fi access at the cost of the employer.
- Offer those with the ability to productively work from home the opportunity to do so without fear of reprisals considering the most vulnerable groups.
- Train all workers in communication platforms such as ZOOM, MS Teams and Google Meet to enable communication with and by them.
- Protect business critical functions such as payroll and accounts departments.
- Split up work teams to avoid everyone becoming ill at the same time.

- Arrange work rosters with registers, screening of everyone upon entry to the workplace, ensure protocols are strictly enforced and all workers who do not pass the initial screening are to be denied access to the work premises and advised to either self-isolate or be referred for testing.
- Reduce contact between leadership teams to ensure operational continuity.
- Move meetings to conference calls or online meetings where possible.
- Obtain all required permits after developing the required plan to demonstrate clearly that the construction site is 'COVID-19 ready'.
- Ensure that the site and all work areas are regularly disinfected and sanitised.
- Actively force social distancing of at least 1.5 metres in all work areas by creating spaces and placing markers on floors.

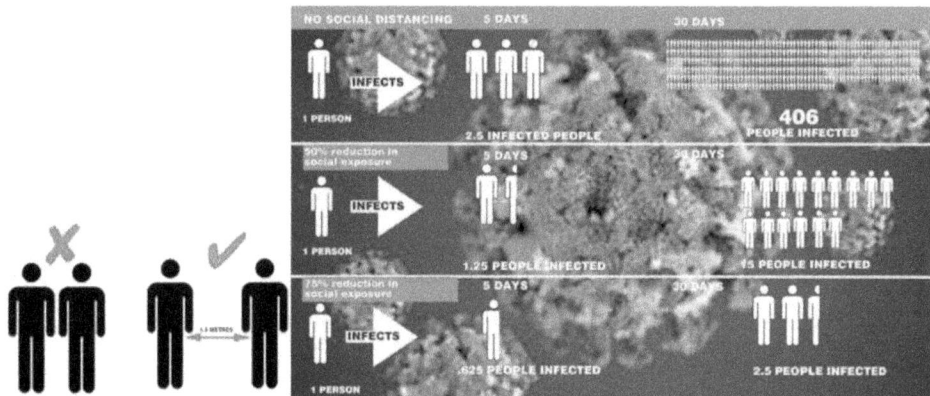

Figure 8.5: The effect of social distancing on infection rate

(Source: https://globalnews.ca/news/6709071/coronavirus-social-distancing-math/)

8.9.1 VACCINES AND VACCINATION

Given the currency of the information on vaccines and vaccinations the following online sources were consulted to provide the latest information and details: https://www.who.int/emergencies/diseases/novel-coronavirus-2019/covid-19-vaccines/html; and https://sacoronavirus.co.za/vaccine-updates/html.

The aim of vaccines and vaccination is to prevent morbidity and mortality. Vaccines work by training and preparing the body's natural defences – the immune system – to recognise and fight off the viruses and bacteria they target. After vaccination, if the body is later exposed to those disease-causing germs, the body is immediately ready to destroy them, preventing illness. As of 18 February 2021, at least seven different vaccines across three platforms have been rolled out in countries. At the same time, as part of a global exercise, more than 200 additional vaccine candidates are in development, of which more than 60 are in clinical development. COVAX is part of the ACT Accelerator, which WHO launched with partners in 2020 and aims to end the acute phase of the COVID-19 pandemic by:

- speeding up the development of safe and effective vaccines against COVID-19
- supporting the building of manufacturing capabilities
- working with governments and manufacturers to ensure fair and equitable allocation of the vaccines for all countries – the only global initiative to do so.

Vaccines are a critical new tool in the battle against COVID-19. Working as quickly as they can, scientists from across the world are collaborating and innovating to produce tests, treatments and vaccines that will collectively save lives and end the current COVID-19 pandemic. The emergency that variants of the virus pose makes the global response challenging since the virus mutates constantly, resulting in the production of new variants with varying impacts on those who are infected.

COVID-19 vaccines go through a rigorous, multi-stage testing process, including large trials that involve tens of thousands of participants. These trials, which include people at high risk for COVID-19, are specifically designed to identify any common side effects or other safety concerns. Once a clinical trial shows that a COVID-19 vaccine is safe and effective, a series of independent reviews of the efficacy and safety evidence is required, including regulatory review and approval in the country where the vaccine is manufactured, before the WHO considers a vaccine product for prequalification. An external panel of experts convened by the WHO analyses the results from clinical trials, along with evidence on the disease, age groups affected, risk factors for disease, and other information. The panel recommends whether and how the vaccines should be used.

Available vaccine details (as at 18 February 2021) based on data released by WHO and the National Institute for Communicable Diseases (NICD) in South Africa are as follows:

Pfizer/BioNTech Vaccine
- Regulatory: Emergency Use Authorizations by US Food and Drug Administration including WHO Prequalification Programme
- Efficacy: > 90% protection – two-dose vaccine
- Rollout has happened in a few countries
- Storage: minus 70°C (limitation for SA as we have limited commercial ultra-low cold chain storage)
- Effective against the 501Y.V2 variant.

AstraZeneca/University of Oxford Vaccine
- Regulatory: Approved as Emergency Use Authorizations by Medicines and Healthcare products Regulatory Agency and Drugs Controller General of India
- Efficacy: 70% efficacy – two-dose vaccine
- AZ has outsourced the production of the vaccine to various sites globally, including the largest vaccine producer globally – the Serum Institute of India (SII)
- This vaccine is likely to be widely used globally due to temperature stability and volumes
- Storage: 2–8°C.

Johnson & Johnson
- Regulatory: Emergency Use Authorisation by FDA
- Single-dose product
- Vaccine has shown to be 66% effective
- Product will also be manufactured at the Aspen facility in South Africa
- Refrigerator storage
- Effective against the 501Y.V2 variant.

Moderna
- Regulatory: Emergency Use Authorizations by FDA
- Two-dose vaccine
- Storage: minus 20 °C
- Effective against the 501Y.V2 variant.

Sputnik V
- In Phase 3 clinical trials in the UAE, India, Venezuela and Belarus
- Sputnik V is already registered in 17 countries
- Sputnik V is a two-dose vaccine
- Efficacy of over 90%
- Storage: The lyophilized vaccine can be stored at a temperature of +2 to +8°C.

CoronaVac (Sinovac)
- In phase three trials in various countries
- Interim data from trials in Turkey and Indonesia show 91.25% and 65.3% effective respectively
- Storage: Refrigerator at 2–8°C.

8.9.2 HERD IMMUNITY

When many people in a community are vaccinated, the pathogen has difficulty circulating because most of the people it encounters are immune. This is called herd immunity. But no single vaccine provides 100% protection, and herd immunity does not provide full protection to those who cannot safely be vaccinated. But with herd immunity, these people will have substantial protection, thanks to those around them being vaccinated. Vaccinating not only protects those vaccinated, but also protects those in the community who are unable to be vaccinated because of their health condition or age, for example.

South Africa is rolling out its national COVID-19 vaccine programme, which aims to vaccinate 40 million South Africans. The South African government is working closely with the South African Health Products Regulatory Authority (SAHPRA) to ensure there is no delay. South Africa's vaccination campaign is guided by science and this means the country may need to change the choice of vaccine it uses. This was demonstrated in the case of the AstraZeneca vaccine, which was found to be less effective against the 501Y.V2 variant while the Johnson & Johnson vaccine proved effective against the variant. While the Johnson & Johnson vaccine may have a lower efficacy rate than other alternatives, the fact that it does not require the same rigorous storage requirements and is a single-dose vaccine makes it the most viable for South Africa.

The vaccine is being administered free of charge at various points of service across all parts of the country. The country's vaccination campaign draws on the principles of universal health coverage where all adults living in South Africa have access to the vaccine. This is the largest vaccination campaign undertaken in the history of the country and stretches across 52 districts and 280 wards to reach 40 million people. The programme entails procurement, distribution, actual vaccination, monitoring, communication and mobilisation.

The vaccination system will be based on a pre-vaccination registration and appointment system at a specific vaccination site. The system will help the government to calculate the number of doses needed at any point in time. All South Africans who are vaccinated will be placed on a national register and provided with a vaccination card. After targeted groups receive the vaccine, mass vaccinations will take place in urban centres at pharmacies, health facilities, community halls and schools. These sites will have to be registered and must comply with several requirements to secure and safeguard the vaccination process.

An electronic vaccination data system (EVDS) will assist with the rollout of COVID-19 vaccines across the country. EVDS is an online self-enrolment portal where South Africans can register via a digital device for an appointment. Those who qualify will be sent a notification through SMS informing them of the time and place that the vaccine will be available. They will have to provide their ID, a contact number and unique code that is sent to them when they are at the vaccination site. Those residents who do not have access to the internet can approach healthcare facilities to assist them with assisted registration on the EVDS.

As the vaccination campaign gains traction with more and more South Africans becoming fully vaccinated some of these requirements will be relaxed over time. Already booster vaccination shots are being administered much the same as the annual flu vaccination shots that have been available for many years already.

Safe and effective vaccines will be a gamechanger: but for the foreseeable future everyone must continue wearing masks, physically distance themselves and avoid crowds even as events are opening up for fully vaccinated spectators and participants. Being vaccinated does not mean that caution can be thrown to the wind because it is still not clear to what degree the vaccines can protect not only against the disease but also against infection and transmission. The emergence of various variants of the virus such as the Delta, Lambda and Omicron variants confirm the need to be cautious and continue to adhere to strict COVID-19 protocols to reduce the likelihood of further waves of infection.

Finally, contractors should continue to monitor information-providing sources such as WHO, the Department of Health, and the NICD for more current information and trends on dealing with COVID-19. They should also consult the provisions of the Disaster Management Act 57 of 2002, as amended, from time to time in terms of section 27(2). COVID-19 will not be the last pandemic or epidemic, considering that it has been the fifth such large-scale health challenge to the world in the last century. Therefore, the lessons learnt from dealing with COVID-19 will be invaluable when faced with similar health challenges on construction sites in the future.

8.10 VIBRATION

Construction sites are renowned for being noisy working environments. Many tools, machines and equipment generate different levels of vibration. Severe vibration exposure occurs in the use of chain saws, jackhammers, vibrators and plate compactors. Percussion drills and percussive tools are the tools that cause the most vibration damage. Transportation systems such as trucks or earth-moving machinery produce tremendous vibration when operated. Vibration causes vascular disorders of the arms and bony changes in the small bones of the wrist. Vascular changes can be detected by X-ray examination of the wrist. The most common findings are rarefaction of the lunate bone.

Whole-body (WBVS) and hand-arm vibrations (HAVS) occur during the use of hydraulic and power tools, and mechanical plant and equipment. HAVS is associated with an increased risk of lower back pain, sciatic pain and degenerative changes in the spinal system including lumbar intervertebral disc disorders. Vascular disorders relative to damage of blood vessels manifest with symptoms such as whiteness of the fingers with numbness and 'pins and needles' and loss of finger dexterity. Neurological disorders resulting from damage to the nervous system manifest in numbness, tingling sensations and loss of hand-grip strength. Other symptoms include pain or stiffness in the hands and lower arms, and a decline in strength and dexterity.

Figure 8.6: Example of a worker using a plate compactor, which generates considerable vibration, on a construction site in KwaZulu-Natal

Table 8.1: Vibration, consequences and recommendations

Whole-body vibration	Consequences	<ul><li>Damage to bones and joints</li><li>Morphological changes in spine</li><li>Digestive system problems</li><li>Female reproductive damage</li><li>Visual impairment</li><li>Defects in vestibular system of ear</li><li>Variations in blood pressure which may lead to heart problems</li></ul>
	Recommendations for prevention	<ul><li>Do not remain on a vibratory surface longer than necessary</li><li>Maintain vibratory machinery regularly to prevent the development of excess vibration</li><li>Keep machine controls remotely located from vibratory surface if possible</li><li>Try to have the source of vibration or the station where workers are stationed vibration-isolated to reduce exposure</li></ul>

Segmental vibration	Consequences	• Contract Vibration White Finger or Raynaud's syndrome • Nerve degeneration that result in the loss of senses of touch and heat • Muscle atrophy and tenosynovitis • Carpal tunnel syndrome
	Recommendations for prevention	• Do not smoke while using the vibratory hand tool (nicotine reduces the blood supply to the hand and fingers) • Let the tool do the work, while you just hold it lightly • Only use the tool when necessary and operate less than full throttle if possible • If you feel signs of tingling, numbness or see white or blue fingers, consult a doctor • Take regular rest breaks to avoid continuous exposure over long periods of time

Vibration can be classified as either (a) whole body vibration that is transmitted through the feet, back or buttocks, or, (b) segmental vibration such as hand-arm transmitted through the hands and arms when working with hand tools. Table 8.1 summarises the health consequences when exposed over lengthy periods to the two different types of vibration, and recommendations for treating each of these. Vibration can be controlled in various ways, and this should be done as best as possible to protect construction workers from the effects of lengthy exposures to extreme vibration:

- Isolating the vibrations by using special mountings that adjust the centre of gravity as low as possible
- Restricting the duration and magnitude of exposure
- Considering working methods that eliminate or reduce exposure such as minimising the transport of goods or materials
- Damping of vibrations, such as wearing padded gloves, or designing or redesigning the tool or machine with damping material between the tool itself and the handle
- Maintaining properly sharpened cutting tools
- Using sandwich structures
- Designing an operating platform remotely from the vibrating equipment or machine
- Warming the operators' hands or handles of the vibratory tools in cold conditions
- Training workers to use their eyes while working and to inspect their hands regularly for injury.

8.11 ALCOHOL AND DRUGS IN THE WORKPLACE

A construction worker may not execute construction activities when under the influence of alcohol or drugs such as dagga (cannabis) or Mandrax. Any worker who uses alcohol or drugs on the construction site while working must face severe disciplinary action such as dismissal or face criminal charges because of the threat they pose to themselves and their fellow workers. When a worker works while intoxicated, he will put himself and his fellow workers at risk, which in itself is very dangerous and could cause an unnecessary and avoidable accident.

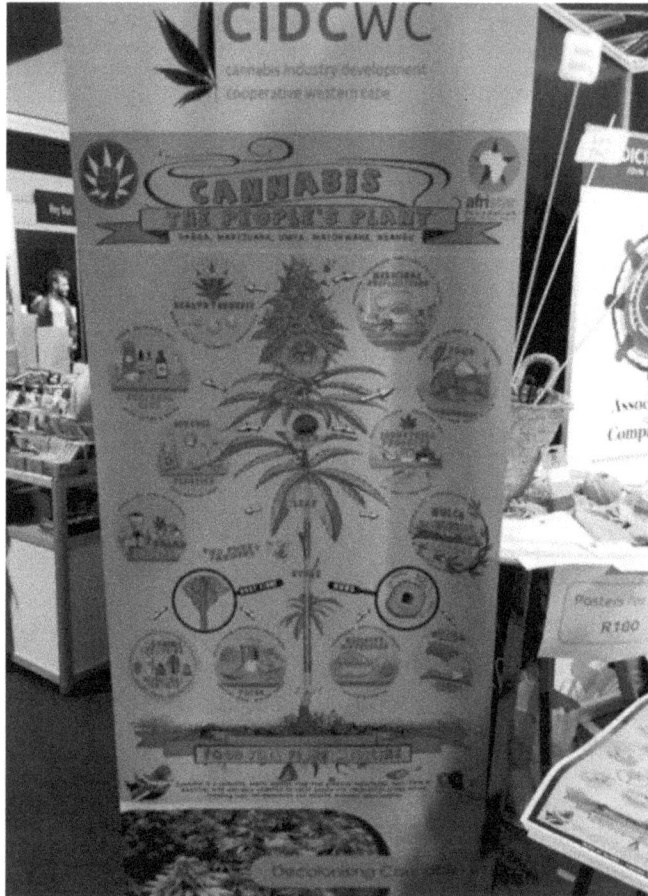

Figure 8.7: A poster at a cannabis expo in Cape Town where cannabis products and technology were being exhibited and marketed

Medication with a high alcohol content, such as cough mixtures, can also cause intoxication and drowsiness. This medication will cause delayed reaction and reduced concentration in the worker taking it. It is very important that construction workers ask the prescribing doctor or pharmacist whether the medication that is being prescribed will have these side or direct effects, or the pamphlet supplied with the medication should be read before it is taken. If the medication will induce delayed reactions, drowsiness and reduced concentration the worker should ask for alternative medication that will not have these side effects. It is very important for workers on medication that could be problematic to inform the supervisor and H&S representative of the medication that is being taken and of its side effects.

8.11.1 IMPACT OF CANNABIS ON CONSTRUCTION SITES

The use and abuse of cannabis by construction workers, given the likelihood that its use will be more visible and brazenly open in South Africa after the court ruling allowing for its limited personal use, will lead to several impacts on construction sites. Construction will be affected unless its use at work or the after-effects from use at home or off-site is controlled on construction sites.

Figure 8.8: A poster in Cape Town where cannabis products such as cannabis-infused alcoholic drinks were being exhibited and marketed

Effects of the use and abuse of cannabis include the following:
- High rates of absenteeism by construction workers who use cannabis and have after-effects from use before coming to work
- Loss of productivity at work because of the lasting effects of use off site
- Violent and unpredictable behaviour that could even lead to petty crimes such as theft and pilfering to fund the cannabis habit
- Steadily decreasing work quality resulting in unnecessary and costly rework to fix poor workmanship and errors
- Increasing inability to pay attention and concentrate for any length of time

- Needless risk taking threatening workplace safety and the safety of fellow construction workers at work
- High labour turnover with the associated recruitment costs.

8.11.2 RECOMMENDATIONS FOR CONTROL OF CANNABIS ON CONSTRUCTION SITES

It is important for the use of cannabis and its after-effects to be effectively managed. Suggestions to improve the control of the use of cannabis itself on site or to reduce the likelihood of construction workers coming onto construction sites with the effects of cannabis use off site include the following:
- The need for management including supervisors and security personnel to be knowledgeable, aware and vigilant
- Create behavioural change on the construction site through the combined use of the traditional extrinsic pathway governed by systems and rules with rewards and punishments and the intrinsic pathway which establishes voluntary compliance via individual commitment to construction H&S
- Development of intensive awareness programmes presented regularly using multiple media as part of wellness programmes
- Implementation of peer interventions and support initiatives since construction workers are more likely to respond to their peers than their supervisors
- Development and communication with the involvement of trade unions and H&S representatives of a 'no or zero tolerance' substance abuse policy that includes regular random testing
- Disciplinary action where the substance abuse policy has been breached by any construction worker
- Improving access and egress control of construction sites that may include random physical body searches and restriction of worker exits during working hours
- Conducting frequent routine inspections of areas of construction work activity and ablution facilities on site
- Setting up an anonymous helpline for workers to report use and to seek counselling if addicted to cannabis
- Consideration for a strict no smoking policy or provision of designated smoking areas that are properly monitored
- Reducing workplace stressors that include adjusting workloads and improving construction site culture and controls through holistic wellness programmes
- Provision of lockers on site for construction workers in properly monitored areas.

8.12 HIV AND AIDS IN THE WORKPLACE

According to the Southern African Development Community (SADC), Southern Africa, with a combined population of only 3.5% of the world's population, accounts for 35% of the people living with HIV and AIDS and about half of all infections in Africa. The high number of the productive adults in the labour force dying from or living with the disease is alarming. Unfortunately, with the shift in focus and attention to COVID-19, HIV and AIDS while being as devastating and deadly has been pushed aside. Further, UNAIDS estimates that over one million people died of AIDS in Southern Africa in 2001, and cumulatively the total number of deaths in the region since the start of the epidemic is over 20 million.

HIV and AIDS, which are not the same thing, threaten productivity, profitability and the welfare of workers and their families. HIV refers to the retrovirus with which construction workers can become infected. Only testing will confirm the serostatus of any worker. On the other hand, AIDS is the illness that manifests in HIV-positive workers after several years for which there is as yet no known cure or vaccine. It is possible for someone who is HIV-positive or sick with AIDS to live a reasonably normal life for many years by using anti-retroviral drugs and treatments. It is only when the C4 white blood corpuscle count reaches very low levels such as around 200 that the construction worker will become extremely ill. Companion afflictions such as TB exacerbate the state of illness. Workplace HIV and AIDS policies and programmes can play a vital role in raising awareness around HIV, preventing HIV infection and caring for people living with HIV.

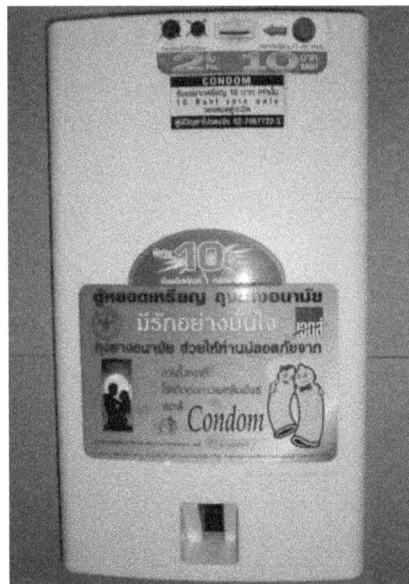

Figure 8.9: Example of a condom dispenser in Bangkok, Thailand

An HIV and AIDS policy, developed and driven by a construction company should involve the following objectives:

- Define the position of a construction company on HIV and AIDS and set out clear guidelines on how HIV and AIDS will be managed on the construction site.
- Align the response on the construction site to the broader legal framework.
- Ensure fairness in the way all construction workers are treated, whether HIV-positive or not.
- Identify and protect the rights and responsibilities of both construction employers and workers in the context of HIV and AIDS.
- Set standards of behaviour expected of all construction employers and their workers.
- Establish consistency within the construction company.
- Set the standard for communication about HIV and AIDS.
- Provide a good foundation upon which to build an effective HIV and AIDS construction workplace programme.
- Inform constructions workers about assistance that is available.
- Indicate commitment to dealing with HIV and AIDS.
- Ensure consistency with national and international practices.

Worker rights

- No worker, or applicant for employment, may be required by their employer to undergo an HIV test. HIV testing by or on behalf of an employer may only take place where the Labour Court has declared such testing to be justifiable in accordance with section 7(2) of the Employment Equity Act.
- All persons with HIV or AIDS have a right to privacy, including privacy concerning their HIV or AIDS status. There is no legal duty on a worker to disclose their HIV status to their employer or to other workers.
- An employer cannot demand to know if the cause of an illness is HIV infection.
- A doctor or health care worker who informs an employer about the HIV status of a worker without their consent is acting against the law and failing to recognise the right of the worker to confidentiality.
- A worker with HIV and AIDS may not be dismissed because he or she is HIV-positive or has AIDS in terms of section 187(1)(*f*) of the Labour Relations Act 66 of 1995. However, where there are valid reasons related to their ability to continue working and fair procedures have been followed, their services may be terminated in accordance with section 188(1)(*a*)(i).
- An employer is obliged to provide, as far as is reasonably practicable, a safe and healthy workplace according to section 8(1) of the Act, which may include ensuring that the risk of occupational exposure to HIV is minimised.
- A worker who is infected with HIV because of an occupational exposure to infected blood or bodily fluids, may apply for benefits in terms of section 22(1) of the Compensation for Occupational Injuries and Diseases Act 130 of 1993.

- In accordance with the Basic Conditions of Employment Act, all workers must receive certain basic standards of employment, including a minimum number of days' sick leave.
- A registered medical aid scheme may not unfairly discriminate directly or indirectly against its members based on their 'state of health' according to section 24(2)(*e*) of the Medical Schemes Act 131 of 1998.

8.13 MANAGING ASBESTOS IN BUILDINGS

Given the severe and terminal effects of lengthy exposure to asbestos fibres or dust the use of asbestos in the construction industry around the world is strictly forbidden. However, many existing older builders were built using asbestos roof sheeting, rainwater goods, lagging and insulation. When these buildings are worked on and the asbestos is disturbed those construction workers could be exposed to the hazards of working with friable and brittle asbestos. Therefore, managing asbestos is necessary to help prevent risk to construction workers or other users of buildings. Asbestos can be found in all pre-2008 constructed non-domestic buildings whatever the type of business.

Inhalation or ingestion of air containing asbestos fibres can lead to asbestos-related diseases such as cancers of the lungs and chest lining. Asbestos is only a risk when asbestos fibres are released into the air and inhaled. Those who are mostly at risk of asbestos fibres are construction workers doing demolition work or maintenance and repairs, roofers, electricians, painters and decorators, joiners, plumbers, gas line installers, plasterers, shopfitters, heating and ventilation engineers, and surveyors. Moreover, anyone dealing with electronics such as telephone services, IT engineers and alarm installers may well be at risk. Only a specialist contractor may be employed to deal with the removal and disposal of asbestos, with each worker wearing special PPE to protect themselves from exposure. There is a long delay between the first exposure to asbestos and the onset of disease and this can vary between 15 and 60 years.

8.13.1 ASBESTOS-RELATED DISEASES

Exposure to asbestos causes diseases such as non-malignant pleural disease, asbestosis, lung cancer and mesothelioma.
- *Non-malignant pleural disease* is the development of diffuse pleural thickening or pleural plaques which may reduce lung volumes.
- *Asbestosis* is a fibrotic interstitial lung infection resulting in shortness of breath on exertion. Dry cough is associated with the later stages of disease. Lung function may become impaired.
- *Lung cancer* is caused by asbestos exposure. Smoking increases the risk of lung cancer. However, this cancer takes many years to develop.
- *Mesothelioma's* early symptoms include weight loss, fever and night sweating. Chest pain, breathlessness on exertion, and/or pleural effusion are present. It may also

result in abdominal discomfort, change in bowel habit and weight loss. It has a long latency period of about 30 years.

8.13.2 TYPES OF ASBESTOS IN BUILDINGS

There are three main types of asbestos found in buildings that are all considered to be carcinogens, namely:
* Blue asbestos – crocidolite
* Brown asbestos – amosite
* White asbestos – chrysotile

Blue and brown asbestos are more dangerous than white forms of asbestos. However blue asbestos, also known as crocidolite, is the most dangerous because its mineral fibres are finer and sharper.

8.13.3 COMMON PLACES WHERE ASBESTOS CAN BE FOUND IN BUILDINGS

Asbestos can be found within many old buildings as it was a commonly used building material for a very long time. The common uses of asbestos include:
* Sprayed asbestos (limpet) (blue or brown) has been used in fire protection ducts and on structural steel members.
* Lagging (blue or brown): > 85% content can be found in thermal insulation of pipes and hot water cylinders.
* Asbestos insulating boards (AIB) (blue or brown): > 85% content has been used in fire protection, thermal insulation, wall partitions, ducts, soffits, ceiling and wall panels.
* Textured coatings containing asbestos have been used in decorative plasters and paints.
* Asbestos cement products – flat and corrugated sheets (approximately 10-15% content) have been used in roofing and wall cladding, gutters, rainwater pipes and water tanks.
* Bitumen or vinyl materials containing asbestos have been used in roofing felts, floor and ceiling tiles.

8.13.4 DUTY TO MANAGE ASBESTOS

It is important and critical for maintenance and facilities managers to manage asbestos in non-domestic buildings. It is also the responsibility of every maintenance manager to ensure the health and safety of occupants or users within the affected buildings. Anyone who is aware of the presence of asbestos within the building must report that to the maintenance manager to ensure the health and safety of everyone who might be or has been exposed. The duties to manage asbestos are listed as follows:
* Finding out if there is asbestos present within the premises, its location and its condition

- Maintaining up-to-date records of location and condition of asbestos on premises
- Assessing the risk from asbestos material
- Preparing a plan that sets out in detail how the risk is going to be managed
- Implementing plan of action
- Reviewing and monitoring the plan and arrangements made to put into action
- Providing information on location and condition to those who might have to work on or disturb asbestos.

8.13.5 STEPS TO CONSIDER WHEN ASSESSING RISK OF ASBESTOS EXPOSURE

The following steps must be considered when assessing potential risks caused by exposure to asbestos in the building:

Step 1: Finding out if asbestos is present within the building

The use of asbestos was prohibited in 2008 in South Africa. In assessing the presence of asbestos within a building, it may be assumed that asbestos is present in all buildings built or refurbished before 2008 unless there is strong evidence against its presence. It may also be necessary to do the following:
- Examine building plans and any other relevant information, such as builders' invoices, which may state if and where asbestos was used in the construction or refurbishment of the premises.
- Carry out a thorough inspection of the premises both inside and out to identify materials that are, or may be, asbestos.
- Consult others, such as the architects, construction workers or H&S representatives, who may be able to provide more information and who have a duty of cooperation to make this available.

Step 2: Assessing the condition of asbestos-containing materials

In assessing the condition of asbestos-containing materials, certain conditions must be considered:
- Whether the surfaces of the materials containing asbestos are damaged, frayed, brittle or scratched
- Whether the surface sealants are peeling or breaking off
- If the materials are becoming detached from their base (particularly within pipes and boiler lagging and sprayed coatings)
- If the protective coverings designed to protect the material are missing or damaged
- Whether there is asbestos dust or debris from damage near the material.

Step 3: Survey and sample for asbestos

The survey of the premises and identification of asbestos in materials must be conducted by a suitably trained person with the necessary special certification and experience, who must provide evidence of liability insurance.

Step 4: Keeping written records

Clear and simple records must be kept and always readily available. The records on asbestos must indicate known or presumed areas where asbestos may be present, type of asbestos, form and condition.

Step 5: Acting on findings

In order to act on the findings, a priority plan must be developed and drawn up. Materials with damage must be given a higher priority to ensure they are repaired, sealed or enclosed by specially trained workers. The following factors must be considered when assessing the likelihood of disturbances of materials containing asbestos:
- Location, amount and condition of the materials containing asbestos
- Position where materials are likely to be disturbed
- Quantity of asbestos
- Accessibility
- If asbestos is closer to normal working areas
- The number of people who use the area where asbestos is present
- If maintenance work, refurbishment is likely to be carried out where there is asbestos.

Step 6: Keeping records up to date

Regular inspection of the condition of materials containing asbestos must be carried out at least every 6 to 12 months and updated in the records. The action plan must also be regularly reviewed and monitored.

8.13.6 CHECKLIST ON MANAGING RISK OF ASBESTOS EXPOSURE

The checklist sets out the necessary steps for managing risk of asbestos:
- Find – check for presence of asbestos.
- Condition – check condition of material.
- Presume – material contains asbestos unless strong evidence it does not.
- Identify – arrange for material to be sampled and identified by a specialist.
- Record – location and condition on a plan or drawing.
- Assess – decide if material is likely to be disturbed.
- Plan – prepare and implement plan to manage risks.

Table 8.2: Managing the risk of asbestos exposure

Minor damage	Good condition
Material should be repaired and/or encapsulated Condition of material should be monitored regularly and if possible labelled Inform contractor and any worker likely to work on or disturb material	Condition of material should be monitored regularly Where possible material should be labelled Inform contractor and any worker likely to work on or disturb material
Poor condition	**Asbestos disturbed**
Asbestos in poor condition should be removed	Asbestos likely to be disturbed should be removed

Health records of every worker exposed to asbestos must be kept for 40 years from date of last entry. Clinical records must also be kept that include a copy of the completed respiratory symptom questionnaire, details of examination findings and any other tests. The statistical records must be made available when requested.

8.14 STRESS AND MENTAL HEALTH

The mental health of the construction workforce is an important health and safety concern for the construction industry. Notably, mental health issues are more prevalent in the construction industry than in other general industries. COVID-19 has exacerbated the challenge due to its impact on the mental health of both management and construction workers brought about by uncertainties, delays, disruptions, changed working conditions and new ways of conducting business. Mental health has been described as being a state of well-being in which every individual realises his or her own potential, can cope with the normal stresses of life, can work productively and fruitfully, and is able to make a contribution to their community. Psychosocial hazards refer to those factors in the design or management of work that increase the risk of work-related stress which can then lead to psychological or physical harm. Individual studies show that work-related psychosocial hazards have negative implications for mental health. Consequently, these hazards cause substantial personal and productivity losses on construction projects.

Stress has been described as the developed body adjustment in response to external threats and can generally be divided into two types, namely physical stress, which refers to physical body adjustments such as breathing difficulties and sleep problems, and emotional stress, which often manifests as depression and feelings of helplessness. Suffering either physical or emotional stress is very detrimental for construction workers as it could lead to morbidity, greatly reduced productivity, a highly increased safety risk and impaired interpersonal relationships. For construction workers, suffering emotional stress symptoms, such as being distressed and sad, could lead to distracted attention at work and in turn cause unintentional breaches of H&S regulations, safe work procedures and construction company H&S policies. Furthermore, it is also possible that those with physical stress symptoms, such as breathing difficulties and sleep problems, may not wear H&S PPE and/or follow proper safe working procedures. According to the Health and Safety Executive (HSE) in the UK stress is the adverse reaction construction workers have to excessive pressures or other types of demand placed on them on construction sites. There is a clear distinction between pressure, which can create a 'buzz' and be a motivating factor, and stress, which can occur when this pressure becomes excessive. Construction-related occupational stress can lead to poor health which may result in injury. About one in five (20%) workers report that they find their work either very or extremely stressful. Over half a million people in the UK report experiencing work-related stress at a level they believe has actually made them ill. Each case of stress-related ill health leads to an average of 29 working days lost.

Because of its dynamic and uncertain characteristics, the construction industry is renowned for being a high-pressure work environment and one of the most stressful occupations. Consequently, construction workers are exposed to high levels of psychosocial hazards and stress. Further, the temporary nature of construction projects makes it difficult to provide permanent positions and long-term support for construction workers. Job insecurity, role conflict, role ambiguity and interpersonal conflict contribute significantly to mental health problems in the construction industry. Additionally, under the pressure of resource-constrained contracts, construction workers must strive hard to complete projects on time, to standard and within budget. Such conditions in the construction industry cultivate psycho-social hazards that adversely affect the mental health of workers, manifesting in anxiety, depression and even suicide. High work demands tend to have more adverse mental health implications. Attention therefore should be given to avoid overburdening construction workers with excessive work demands, so as to reduce the negative influence on mental health and related poor job performance.

Causes of occupational stress

Stress sets off an alarm in the brain, and the brain responds by preparing the body for a defensive action. This biological response is sometimes called the fight or flight response and is important because it helps the body cope with threatening situations. The nervous system is aroused by this action and releases hormones to heighten the senses, increase heart rate, deepen respiration and tense the muscles.

Construction-related occupational stress results from the interaction of construction workers with the working environment. However, differences in individual characteristics such as personality and coping style have an impact on predicting whether certain job demands will result in stress. For example, what is stressful for one worker may not necessarily be problematic to another worker. Therefore, it is generally accepted that working conditions and the working environment play a vital role in causing job stress. It has become evident that stress causes several types of chronic illnesses such as cardiovascular disease, musculoskeletal disorders and psychological disorders. Furthermore, stress-related syndromes often manifest as psychosomatic symptoms.

Early symptoms of occupational stress include, for example:
* Headache
* Insomnia
* Short temper
* Upset stomach
* Job dissatisfaction
* Low morale
* Low energy levels
* Weak immune system.

NIOSH Model of Job Stress

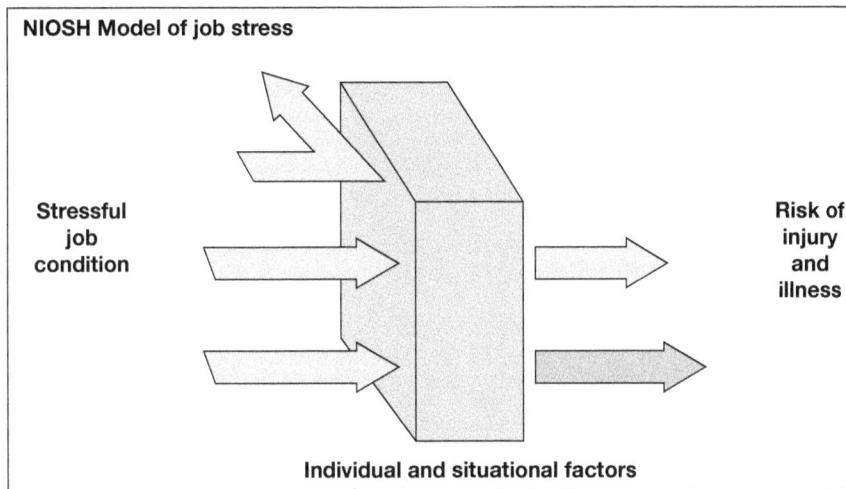

Figure 8.10: National Institute for Occupational Safety and Health (NIOSH) Model of Job Stress

Planned personal contacts

For effective communication to be pervasive throughout the organisation, managers at all levels should have planned individual contact sessions with each construction worker who reports to them. While group meetings have their role, it is likely that not all construction workers will speak up and/or air their views, concerns and misgivings. Individual contact sessions fill this gap.

By keeping an open line of communication between construction workers and themselves, managers are in a better position to receive valuable feedback from their workers without having to pursue it. Arguably, planned individual contacts allow for:
* Personalising critical aspects of construction health and safety for each construction worker
* Building better construction health and safety awareness and attitudes on the construction site

- Demonstrating practically and tangibly to each construction worker that the manager has a personal concern for proper working practices and conditions on the project
- Improving supervisor–construction worker relationships.

A planned personal contact is an intentional get-together of a supervisor and a construction worker to discuss a critical construction H&S topic related to the construction work activities that the worker is performing or about to perform. In this way, there is a specific topic for a specific construction worker executing a specific construction activity in a specific situation on the construction site.

It is important that these personal sessions should be frequent enough to influence the attitudes, knowledge and/or skills of the construction worker. At least one personal contact session should be planned with each construction worker each month. Given the challenges of COVID-19 and its after-effects the session does not necessarily have to be in-person, which is preferable, but can be arranged making use of alternative communication platforms that include GoogleMeet, Hangout, Twitter, Instagram, Teams and Zoom.

Management of stress in the workplace

The Management Standards propagated by the HSE in the UK as a tool to manage stress in the workplace cover six key areas of work design that, if not properly managed, are associated with poor health and well-being, lower productivity and increased sickness absence. In other words, the six Management Standards cover the primary sources of stress at work. These are:

1. **Demands** – such as workload, work patterns and the work environment
2. **Control** – such as how much say the worker has in the way they do their work
3. **Support** – such as the encouragement, sponsorship and resources provided by the organisation, line management and colleagues
4. **Relationships** – such as promoting positive working to avoid conflict and dealing with unacceptable behaviour
5. **Role** – such as whether workers understand their role within the organisation and whether the organisation ensures that they do not have conflicting roles
6. **Change** – such as how organisational change (large or small) is managed and communicated in the organisation.

Construction employers have a duty to ensure that risks arising from construction activities are properly controlled. The Management Standards approach helps contractors work with their workers and representatives to undertake risk assessments for stress. The balance between work and family or personal life could further alleviate stress and workers could further seek a support network of friends and co-workers and thus approach work in a relaxed and positive manner.

REVIEW QUESTIONS

1. What are the challenges for the development of occupational health and safety?
2. How can work affect health and health affect work? Give practical examples.
3. Why is a focus on occupational health necessary?
4. What are the five legs of the South African occupational health model?
5. What rights do workers have pertaining to HIV and AIDS?
6. What steps must be considered when assessing potential risks caused by exposure to asbestos on a construction project involving renovations to a pre-2008 building?
7. What threats do migrant workers present on a construction site?
8. What actions should an employer take to deal with the threat of COVID-19 on a construction site?
9. What are the consequences of avoiding a focus on vibration because of construction work activities?
10. What steps should be taken to prevent the use of cannabis and its effects on workers at work?
11. How would ergonomics contribute to improving worker productivity and efficiency?
12. Develop a safe work or operating procedure for a construction site to ensure that it is 'COVID-19 ready', considering all information and regulations issued by the World Health Organization, Department of Health, Department of Labour and National Institute for Communicable Diseases.

REFERENCES

Abedi, M. 2020. A look at the math behind social distancing amid coronavirus. Global News. Available at: https://globalnews.ca/news/6709071/coronavirus-social-distancing-math/

Amponsah-Tawiah, K & Mensah, J. 2016. Occupational health and safety and organizational commitment: Evidence from the Ghanaian mining industry. *Safety and Health at Work*, 7(3): 225–230

Attridge, M, Farley, T, Marion, D & Cycenas, J. 2020. EAP and COVID-19: Employee Assistance Programs (EAP) in the COVID-19 Era: Support for Substance Abuse

Bovenzi, M & Hulshof, C. 1999. An updated review of epidemiologic studies on the relationship between exposure to whole-body vibration and low back pain (1986–1997). *International Archives of Occupational and Environmental Health*, 2(6): 351–365

British Lung Foundation. 2008. Construction tradespeople, Asbestos awareness survey. Available at http://a-a-s-c.org.uk/wp-content/uploads/2012/05/Tradespeople-Asbestos-Awareness-Survey-Feb-2008.pdf

British Occupational Hygiene Society. nd. BOHS Publications - British Occupational Hygiene Society (BOHS). [online]. British Occupational Hygiene Society (BOHS). Available at: http://www.bohs.org/about-us/bohs-publications/

Burdorf, A, Porru, F & Rugulies, R. 2020. The COVID-19 (Coronavirus) pandemic: Consequences for occupational health. *Scandinavian Journal of Work, Environment & Health*, 46(3): 229–230

Campbell, F. 2006. *Occupational Stress in the Construction Industry*. London: The Chartered Institute of Building. Available at: http://www.ciob.org/filegrab/stress.pdf

Canada. nd. *Health Effects of Cannabis*. Available at: https://www.canada.ca/content/dam/hc-sc/documents/services/campaigns/27-16-1808-Factsheet-Health-Effects-eng-web.pdf

CBS News. 2018. South Africa rules marijuana legal in private for adults. https://www.cbsnews.com/news/south-africa-rules-marijuana-legal-in-private-for-adults/

Darnton, A, McElvenny, D & Hodgson, J. 2006. Estimating the number of asbestos-related lung cancer deaths in Great Britain from 1980 to 2000. *Annals of Occupational Hygiene*, 50(1): 29–38

Department of Health, Social Services and Public Safety. 2014. *Workplace Drugs and Alcohol Policies*. Available at: http://www.dhsspsni.gov.uk/stats-workplace

Department of Labour. 1993. Compensation for Occupational Injuries and Diseases Act 103 of 1993

Department of Labour. 1993. Occupational Health and Safety Act 85 of 1993

Department of Labour. 1995. Labour Relations Act 66 of 1995

Department of Labour. 1997. Basic Conditions of Employment Act 75 of 1997

Department of Labour. 1998. Employment Equity Act 55 of 1998

Ethiopia Public Health Training Initiative. 2006. *Occupational Health and Safety*. Available at: https://www.cartercenter.org/resources/pdfs/health/ephti/library/lecture_notes/env_occupational_health_students/ln_occ_health_safety_final.pdf

European Union. 2020. *Q & A on COVID-19*. [online] European Centre for Disease Prevention and Control. Available at: https://www.ecdc.europa.eu/en/novel-coronavirus-china/questions-answers

Goelzer, B. nd. Chapter 30 – Occupational Hygiene. [online] Ilocis.org. Available at: http://www.ilocis.org/documents/chpt30e.htm

Goetsch, DL. 2011. *Construction Safety and Health*. Upper Saddle River, New Jersey: Pearson Education Inc

Hale, A. 2019. From national to European frameworks for understanding the role of occupational health and safety (OHS) specialists. *Safety Science*, 115: 435–445

Harinarain, N & Haupt, T. 2010. Impact of workplace HIV and AIDS polices on stigma and discrimination. In Haupt, TC (ed) The Built Environment 5, Proceedings of the 5th Built Environment Conference, Durban, 18-20 July, pp 453-465

Harinarain, N & Haupt, TC. 2011a. HIV and AIDS: The KZN construction industry response. *Journal of Construction*, 4(2): 13–17

Harinarain, N & Haupt, T. 2011b. Implications of ignoring HIV and AIDS by the construction industry: The South African experience. CIB W99 Conference: Prevention: The Means to the End of Construction Injuries, Illnesses and Fatalities, Washington, D.C., August 24-26

Harinarain, N & Haupt, T. 2012. Threats to effective HIV and AIDS management in construction: Lessons from literature. International Cost Engineering Council – 8th ICEC World Congress 'Quest For Quality: Professionalism in Practice'. International Convention Centre: Durban, South Africa: 23–27 June 2012

Harinarain, N & Haupt, T. 2014b. The lack of management commitment to HIV and AIDS in South African construction. Proceedings of Association of Schools of Construction of Southern Africa (ASOCSA) 8th Built Environment Conference, 28–29 July, University of Kwa Zulu Natal, Durban, South Africa, pp 289–300

Harinarain, N & Haupt, T. 2014c. The vulnerability of the construction industry to HIV and AIDS. *Journal of Construction*, 7(2): 61–66

Harinarain, N & Haupt, TC. 2014a. Drivers for the effective management of HIV and AIDS in the South African construction industry—a Delphi study. *African Journal of AIDS Research*, 13(3): 291–303

Haslam, C & Mallon, K. 2003. Post-traumatic stress symptoms among firefighters. *Work & Stress*, 17: 277–285

Haupt, T. 2019a. An appraisal of the use of cannabis on construction sites. *Acta Structilia*, 26(1): 148–166

Haupt, TC. 2019b. The use of cannabis on construction sites: A review. *Journal of Construction*, 11(3): 29–35

Haupt, TC. 2021. *Management of Safety, Health and Environment in South Africa: A Handbook*. Newcastle upon Tyne: Cambridge Scholar Publishing

Haupt, TC, Akinlolu, M & Raliile, MT. 2019. The use and effect of cannabis among construction workers in South Africa: A pilot study. 10th West Africa Built Environment Research Conference (WABER), Accra, Ghana, pp 1126–1137

Haupt, TC, Deacon, C & Smallwood, JJ. 2005. Respiratory and skin infections in older construction workers. *Occupational Health Southern Africa*, 10(6): 4–9

Haupt, T, Smallwood, J, Kalindindi, S & Varghese, K. 2010. The health and wellbeing of Indian construction workers: A comparison between older and younger. In Barrett, Amaratunga, Haigh, Keraminiyage & Pathirage (eds) Building a Better World: Proceedings of CIB World Congress 2010, Manchester, U.K., May 10-13, 2010, paper 1634

Health and Safety Authority. nd. *New Asbestos Guidelines*. Available at: https://www.hsa.ie/eng/Your_Industry/Chemicals/Legislation_Enforcement/Asbestos/newasbestosguidelines.pdf

Health and Safety Executive (HSE). nd. *Asbestos - Health and Safety in the Workplace*. http://www.hse.gov.uk/asbestos Google Scholar

Health and Safety Executive (HSE) (nd). Occupational Hygiene - HSL. [online] Hsl.gov.uk. Available at: https://www.hsl.gov.uk/what-we-do/occupational-hygiene.

Health and Safety Executive (HSE) (nd). Work-related stress and how to tackle it. Available at: https://www.hse.gov.uk/stress/what-to-do.htm

Health and Safety Executive (HSE). 2002. *Respirable Crystalline Silica, Phase 1*. Available at: http://www.hse.gov.uk/pubns/priced/eh75-4.pdf.

Health and Safety Executive (HSE). 2014a. *Statistics, Mesothelioma.* Available at: http://www.hse.gov.uk/Statistics/causdis/mesothelioma/index.htm

Health and Safety Executive (HSE). 2014b). Musculoskeletal Disorders in Great Britain. Available at: http://www.hse.gov.uk/statistics/causdis/musculoskeletal/msd.pdf

Health and Safety Executive for Northern Ireland. 2009. *Construction Workers: Advice on Cancer Prevention.* Available at: http://www.hseni.gov.uk/advice_on_cancer-2.pdf

Holt, Allan St John. 2001. *Principles of Construction Safety.* Osney Mead, Oxford: Blackwell Science Ltd

International Labour Organisation (ILO). 1985. Convention C161 - Occupational Health Services Convention, 1985 (No. 161). [online] Ilo.org. Available at: http://www.ilo.org/dyn/normlex/en/f?p=NORMLEXPUB:12100:0::NO::P12100_INSTRUMENT_ID:312306

International Labour Organisation (ILO). 1985. Recommendation R171 - Occupational Health Services Recommendation, 1985 (No. 171). [online] Ilo.org. Available at: https://www.ilo.org/dyn/normlex/en/f?p=NORMLEXPUB:12100:0::NO::P12100_ILO_CODE:R171

International Labour Organisation (ILO). 2013. *The Prevention of Occupational Diseases.* Available at: https://www.ilo.org/safework/info/WCMS_208226/lang--en/index.htm

Levy, B & Wegman, D. 2000. *Occupational Health: Recognizing and Preventing Work-related Disease and Injury.* Philadelphia: Lippincott Williams and Wilkins

Liang, Q, Zhou, Z, Ye, F & Shen, L. 2022. Unveiling the mechanism of construction workers' unsafe behaviors from an occupational stress perspective: A qualitative and quantitative examination of a stress–cognition–safety model. *Safety Science,* 145. https://doi.org/10.1016/j.ssci.2021.105486

McAleenan, C & Oloke, D. 2015. *ICE Manual of Health and Safety in Construction.* 2nd ed. London: ICE Publishing

National Institute for Occupational Health. 2020. *Guidelines on Routine and Deep Cleaning in the Workplace.* Available at: https://www.nioh.ac.za/wp-content/uploads/2020/07/Guidance-on-routine-deep-cleaning_1-July-Fin.pdf

National Institute for Occupational Safety and Health. 1999. *STRESS…At Work* (NIOSH Publication No. 99-101). Available at: http://www.cdc.gov/niosh/docs/99-101/

Occupational Safety and Health Administration. 2014. Safety and Health Topics Page, Welding, Cutting, and Brazing 2014

Peto, J, Rake, C, Gilham, C & Hatch, J. 2009. Occupational, domestic and environmental mesothelioma risks in Britain. Available at: http://www.hse.gov.uk/research/rrpdf/rr696.pdf

Prall, J & Ross, M. 2019. The management of work-related musculoskeletal injuries in an occupational health setting: The role of the physical therapist. *Journal of Exercise Rehabilitation,* 15(2): 193

Reese, CD. 2018. *Occupational Health and Safety Management: A Practical Approach.* CRC Press

Richmond, MK, Pampel, FC, Wood, RC & Nunes, AP. 2017. The impact of employee assistance services on workplace outcomes: Results of a prospective, quasi-experimental study. *Journal of Occupational Health Psychology*, 22(2): 170

Rom, W, Hammar, S, Rusch, V, Dodson, R & Hoffman, S. 2001. Malignant mesothelioma from neighbourhood exposure to anthrophylite asbestos. *American Journal of Industrial Medicine*, 40: 211–214

Samuels, W, Haupt, TC & Shakantu, WMW. 2007. An exploratory examination of poor construction practices: Opportunities for ergonomic interventions in the Western Cape. *Occupational Health Southern Africa*, 13(2): 19–25

Smallwood, JJ & Haupt, TC. 2007. Construction ergonomics: An Indian and South African comparison. *Ergonomics SA - Journal of the Ergonomics Society of South Africa*, 19(1): 30–43

Smallwood, J & Haupt, T. 2009. Construction ergonomics: Perspectives of female and male production workers. In Dainty A (ed) Proceedings of 25th Annual Conference of Association of Researchers in Construction Management (ARCOM), Nottingham, U.K., 7–9 September, pp 1263–1272

Smallwood, JJ & Haupt, TC. 2009. Construction ergonomics: Perspectives of female and male production workers. Paper 1BU0021, Proceedings of 17th World Congress on Ergonomics, Beijing, China, August 9-14 (CD-ROM)

Sun, C, Hon, C, Way, K, Jimmieson, N & Xia, B. 2022. The relationship between psychosocial hazards and mental health in the construction industry: A meta-analysis. *Safety Science*, 145. Available at: https://doi.org/10.1016/j.ssci.2021.105485

Tyrer, J. 2015. A new means of quantifying laser safety hazards: The laser micro-mort. *International Laser Safety Conference*

Ugale, C. 2009. *Relationship between Occupational Stress & Job Satisfaction: A Case Study of InfoTekNetAlia*. Unpublished MBA dissertation, University of Wales

Vincent, JH. (2005). Graduate education in occupational hygiene: A rational framework. *The Annals of Occupational Hygiene*, 49(8): 649–659

Wilson, J. 2014. Fundamentals of systems ergonomics/human factors. *Applied Ergonomics*, 45(1): 5–13. Available at: https://doi.org/10.1016/j.apergo.2013.03.021

World Health Organization (WHO). 1986. Constitution: Basic documents. Geneva

World Health Organization (WHO). 1994. Global strategy on occupational health for all: The way to health at work. Geneva

World Health Organization (WHO). 2006. Declaration of workers health. WHO Collaborating Centres of Occupational Health: Stresa, Italy

World Health Organization (WHO). 2010. Healthy workplaces: A model for action. Geneva

World Health Organization (WHO). 2020a. Coronavirus. [online] Who.int. Available at: https://www.who.int/health-topics/coronavirus

World Health Organization (WHO). 2020b. HIV/AIDS. [online] Who.int. Available at: https://www.who.int/news-room/fact-sheets/detail/hiv-aids

Zarate, P, Cuellar, D, Velazquez, L & Cura, L. 2018. 1017 occupational health nurses working as worksite health promotion agents

CHAPTER 9
HAZARD IDENTIFICATION, RISK ASSESSMENT AND MITIGATION

9.1 INTRODUCTION

The construction industry is regarded as one of the most hazardous and risky industries when compared to other industrial sectors. Construction workers are exposed daily to hazards that could have been mitigated to avoid the possible negative outcomes of such exposures. Construction risk is pervasive and can either be considered consciously or most of the time subconsciously. Risk can be perceived as situations in which a decision is made, the consequences of which depend on the outcomes of future events having known probabilities and previous experiences.

One can also suffer from risk homeostasis in terms of which the reduction or elimination of risk will increase other risks or likely acceptance of other risks in return; or the more certain risks are mitigated the more blind one becomes and willing to accept other risks. For example, if one wears a seatbelt to prevent flying through the windscreen in the event of an accident, one could conceivably be willing to drive faster as a result, which in itself would be another risk.

Construction H&S risk management is a management function with the objective of protecting workers and other project stakeholders, assets, environment, and income by avoiding or minimising the potential for loss from risk exposures, and the provision of funds to recover these losses.

9.2 OBJECTIVE OF CONSTRUCTION HEALTH AND SAFETY RISK MANAGEMENT

The objective of construction project H&S risk management includes the following:
* Recognition that risks are difficult to eradicate
* Avoidance of downside risks (risk avoidance)
* Exploitation of opportunities to create construction workplaces that do not threaten the health and safety of construction workers and other project stakeholders
* Making decisions against a predetermined set of construction H&S objectives, rules and priorities based on relevant knowledge, data and information.

9.3 DEFINITION OF RISK

Risk is the presence of uncertainty and it is measured as the deviation from the expected outcome of a given situation or event. Two aspects of risks are considered when defining a risk, namely frequency and severity. A pure risk is a risk that results in loss, damage, disruption or injury. Pure risk can usually be insured against. Risk management is primarily concerned with pure risk. Examples are fire, theft, floods, incidents, accidents and diseases arising from construction activities.

9.4 DEFINITION OF FREQUENCY

Frequency is the number of times the event or exposure occurs over a given period. One could ask these questions: How often will or did it happen? What are the chances or probability of it happening on this construction project and site?

9.5 DEFINITION OF SEVERITY

Severity represents the cost of the damage, consequential loss or injury. Here one could ask the questions: How serious or severe is it? What is the extent of the impact or consequences should the exposure occur?

9.6 CLASSIFICATION OF RISK

There are many ways to classify risk. For example, universal categories are shown in Figure 9.1.

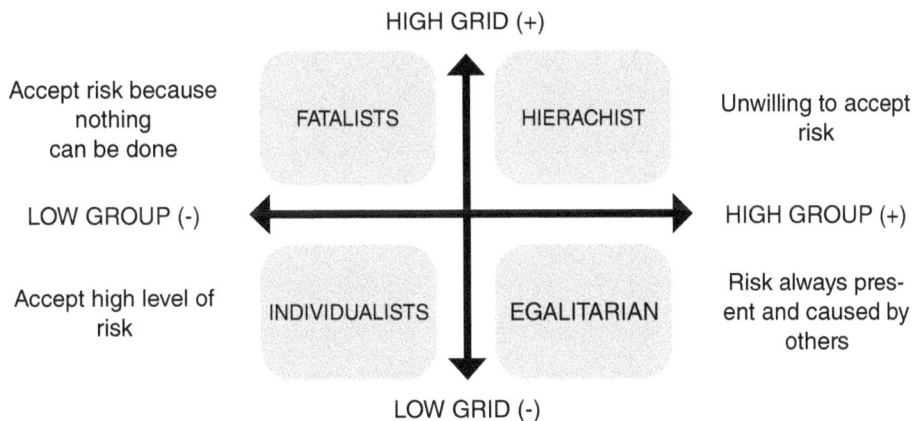

Figure 9.1: Universal categories of risk

Risks may also be classified either as pure risks (insurable risks) or speculative risks. Risks may also be classified as:

- Known risks – there are minor variations in the levels of exposure
- Known unknown risks – the occurrence or exposure is predictable
- Unknown unknown risks – risks about which the company does not even know that they do not know.

Finally, risks may also be classified according to the level of probability and the consequences of impact as shown in Figure 9.2. As the probability increases the more the exposure to the hazard under investigation will be expected. On the other hand, as the impact increases the greater the need for risk management. Where both the impact and probability are high the need for some form of risk management is an imperative and not debatable.

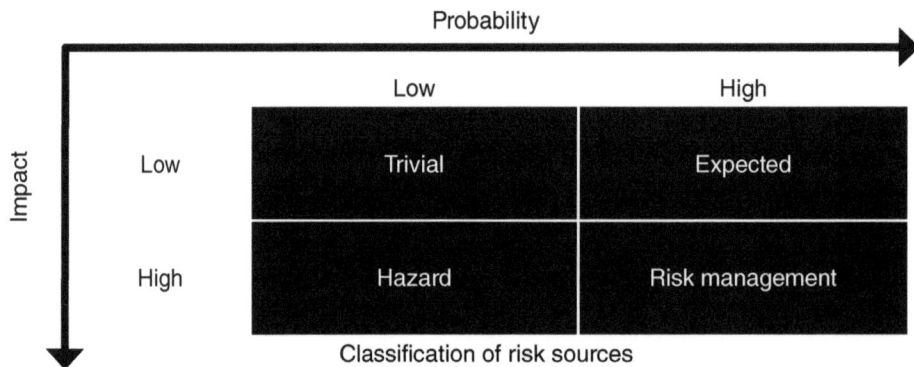

Figure 9.2: Risk classification according to probability and impact

9.7 BENEFITS OF RISK MANAGEMENT

Some of the benefits of risk management include:

- Clarification of project H&S issues from the start of the project
- Promotes uninterrupted progression of activities since all exposures to hazards have been considered
- Promotes communication between all project stakeholders
- Decisions that are supported by thorough analysis
- Instils confidence in the project, third parties and project team
- Clearer understanding of specific project H&S risks
- Build-up of historical data for future use
- Promotes project success because all eventualities have been considered.

9.8 CAUSES OF RISK MANAGEMENT FAILURES

Not all H&S risk management strategies are successful. The failure of H&S risk management on construction projects can be caused by, for example:

- human error
- irrationality
- cultural misunderstandings
- failed communications
- systemic failure
- homeostatic need.

9.9 DISTRIBUTION OF CONSTRUCTION HEALTH AND SAFETY RISK

Generally, construction H&S risk is not distributed evenly across any project. Typically, the risk exposure is greatest during the early stages of the construction project and decreases over time. Figure 9.3 illustrates this trend. There also tends to be a spike as the project draws to a close.

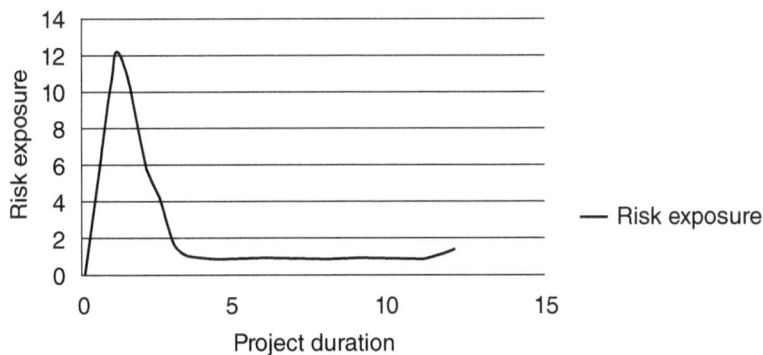

Figure 9.3: Distribution of risk exposure across a project

9.10 RISK-RELATED CONCEPTS

Hazard

A hazard is a condition, activity, object or substance that creates, introduces or increases the frequency or severity of a loss-producing event on a construction project. Hazards are anything that may cause a loss-producing event.

Peril

A peril describes the risk that is or could be involved. Perils are the source of the loss. (Flammable liquid as a hazard could be the cause of a fire (peril) which constitutes a risk of explosion.)

Probability

Probability is an indication of the likelihood that a certain event or exposure could occur in the long run.

9.11 HAZARD IDENTIFICATION, RISK ASSESSMENT AND CONTROLS

Each organisation should have in place procedures for ongoing hazard identification, risk assessment and necessary controls. Hazard identification and risk assessment need to be proactive instead of reactive methodologies. Risks need to be identified, prioritised and documented together with the appropriate controls and interventions. The hierarchy of risk reduction always needs to be implemented, namely:
- Elimination
- Substitution
- Engineering controls
- Signage, warnings and/or administrative controls
- Personal protective equipment.

9.11.1 RISK IDENTIFICATION METHODS

There are multiple approaches that can be used to identify risk. It is important for each construction organisation to decide which approach or method works for them depending on the nature and scale of its operations. The aim is to identify, analyse and control hazardous situations before these situations turn into negative incidents. The earlier the construction H&S personnel become involved, the greater the chances are of identifying the risks involved in a project. This intervention/precaution reduces the probability of occurrence and the severity of the consequences.

9.11.2 RISK IDENTIFICATION TECHNIQUES

Physical inspection

Physical inspection of the construction site before actual work commences can help to identify hazards, which might pose a threat to health, safety and the environment.

Interviews

Interviews should take place during the site inspection with the project team to determine the levels of health, safety and environmental exposures.

Documentary information

The study of documentary information includes, for example, statutory records, process descriptions, logbooks, standards and codes of practice, organisational charts, flow charts, and information from previous construction projects. Other scientifically proven techniques are hazard and operation studies, failure mode and effect analysis, fault tree analysis and hazard indices.

9.11.3 HAZARD IDENTIFICATION AND RISK ASSESSMENT

The Hazard Identification and Risk Assessment (HIRA) should be conducted to ensure the reduction or elimination of risk to those who will be involved in any stage of a construction project, or installation or during the maintenance phase of a project. The process does not need to be precise but should deal with all known hazards, issues or risks where possible, or reasonable. A competent person is required to conduct the HIRA, and a construction H&S specialist could be used to assist in this process. However, the HIRA process must commence during the initiation phase, and be addressed during all the other project phases. For example, in the case of a major installation in the plant, the design related aspects could include, inter alia, concept design in addition to detailed design, while evolving details, schedule, method of fixing and specifications.

STEPS TO BE FOLLOWED

Steps may vary slightly depending on the type of risk assessment used. It is advisable to assemble a risk assessment team given the subjective nature of the process. However, the risks need to be determined from the initial concept or decision to proceed, followed by the drawings where required such as in the case of a major installation, and the work divided into the various activities followed. Thought must be given to the method of procurement, schedule, installation processes to be followed, and products used to limit exposure.

Material safety data sheets (MSDSs) should be obtained from suppliers to determine short- or long-term health risks to workers, whereafter it is possible to determine if alternative products should rather be used. For example, have solvent-based paints been specified for use in a closed environment where ventilation is likely not to be optimum? In such a case, consideration needs to be given as to whether a water-based paint could be used, as PPE may only be used as a last resort, and where no other safer or healthier option is available.

Previous knowledge is useful where the construction work to be done on the project is similar. However, it must be remembered that no two processes, projects, installations or maintenance operations are alike. For example, the placing of storm water culverts in position invariably entails the same procedure. However, the environmental factors such as ground (need to shore) and surrounding traffic (national road or farm area), may alter the risks significantly. Installation of roof sheeting will depend on the pitch of the roof and the size of the sheets. Workers need to be protected from falling, and attachments for fall arrest lines may be required.

Since the HIRA is required to be part of a construction H&S specification and plan it must be recorded. Furthermore, the Occupational Health and Safety Inspectorate of the Department of Labour may require the responsible persons or department in the construction company to indicate endeavours made to identify hazards and mitigate risks particularly if an accident that results in either fatalities or injuries should occur.

A quantitative process is recommended, and risks should be graded from low to extreme. It must be remembered that the medium to extreme risk areas must be addressed. The HIRA needs to be made available to all parties involved and such HIRAs should be kept on file. Should it not be possible to eliminate or mitigate the hazard, the requisite insert should be made in the relevant documentation and, where appropriate, the financial implications should be shown.

RISK QUANTIFICATION

To make an informed decision with respect to deciding, for example, whether a design, detail, method of fixing or specification is appropriate, the risk needs to be quantified. Based upon the principle of reasonably practicable, which includes cost and other goals, a decision can be made to amend or leave the design or other relevant aspects unchanged. Furthermore, a precise estimate is not required, as doing so would be too time consuming, and often there is a lack of data.

Loss frequency is a direct measure of the frequency at which a particular loss occurs over a given period. All incidents must be reported even if no loss occurred. An increase in minor injuries should be taken as an indication of a deteriorating situation. This is an important database for risk evaluation. Investigations in many organisations over a long period of time have indicated the relationship.

It is important to bear in mind that the extent of a loss is not limited to the cost to replace or repair. The immediate or obvious loss or damage may be only 'the tip of the iceberg'. It is accepted in many organisations to adopt and follow the 'Pareto Principle'.

In simple terms, this rule can be defined as follows: 80% of the losses will normally be associated with 20% of the exposure, or as the extent of the loss doubles, the claims frequency will halve.

Other techniques such as statistical estimates, asset values and impact studies are also available to determine possible loss extent.

Figure 9.4 illustrates how the risk can be determined by plotting the likelihood of the risk occurring (probability/frequency) on the vertical axis and the likely severity (consequences/impact) on the horizontal axis. The scores relative to probability and consequences/impact are as follows:
- Likely severity (consequences/impact) – three categories:
  - High – fatality, major injury or illness causing long-term disability
  - Medium – injury or illness causing short-term disability
  - Low – other injury or illness
- Likelihood of occurrence (probability) – three categories:
  - High – certain or near certain to occur
  - Medium – reasonably likely to occur
  - Low – very seldom or never occurs.

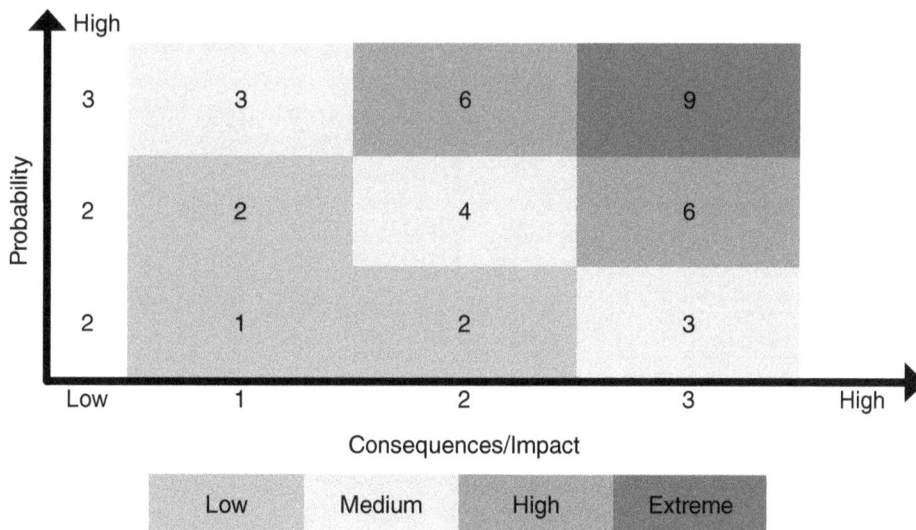

Figure 9.4: Quantifying magnitude of risk using probability/impact matrix

A score of '9' would require elimination of the cause of the problem, for example, selection of a flat roof in lieu of a steep pitched roof structure. A '6' would require elimination, or at least substitution, for example, water in lieu of a solvent-based spray on paint finish. A '4' would require elimination, substitution, or at least containment, for example, screening off areas in which stripping of asbestos-containing materials will be required, or the application of a decorative spray-on solvent-based paint finish. A '3' would require elimination, substitution, containmen or at least an engineering intervention, which often would also require work in accordance with a safe work or operating procedure (SWP/SOP), which would also entail the wearing of personal protective equipment (PPE); for example, a balustrade wall to a flat roof. A '2' would require work in accordance with a SWP/SOP, and entail the wearing of PPE, for example, construction of the balustrade wall in concrete, should that be the solution relative to the '3' score. A '1' would require no intervention. The HIRA needs to be communicated and reviewed together with all identified controls. The final step in the HIRA is to evaluate and prioritise the risks as all risks, their loss frequency and consequences are not the same.

9.11.4 RISK EVALUATION OR QUANTIFICATION

The aim of risk evaluation is to measure the magnitude of one risk relative to another. The aim is also to establish the impact (financially and otherwise) that the risk would have on the organisation. There are multiple approaches to evaluation and quantification of risks. The method used is not important. What is important are the decisions that flow from the process.

It is also important to note that generally there are two types of personalities that will determine the level of risk acceptance and strategy to be adopted. These are risk takers and risk avoiders.

Risk taker	Risk avoider
• Risk lovers • Entrepreneurs and investors • Accept higher exposure – higher variability • Underrate risk	• Risk averse • Low-paid but safe jobs • Overrate risk

Figure 9.5: Risk takers vs risk avoiders

Typically, risk takers will adopt a more lenient approach to hazard identification and eventual risk assessment and quantification. On the other hand, risk avoiders will take a much more conservative approach. For this reason, it is preferable where possible for hazard identification, risk assessment and quantification to be done collectively by more than one person to ensure a balanced approach to the process. This process is shown in Figure 9.6. Note that the various steps in the risk control process are collectively iterative.

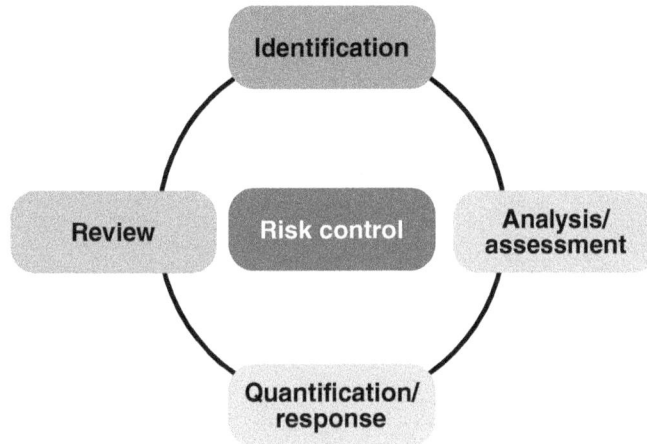

Figure 9.6: Steps in the risk control process

9.11.5 RISK CONTROL TECHNIQUES AND OPTIONS

Generally, there are four main techniques to control construction H&S risk:
1. Risk avoidance – not to take on a risk in the first place
2. Risk elimination – eliminating an existing risk
3. Risk reduction – reducing the frequency and/or the severity of a risk
4. Risk transfer – transferring the risk to someone else or to another place where it poses a smaller risk for the organisation; sub-contracting is a means of transferring the risk of executing a section of the construction work.

These techniques are shown in Figure 9.7.

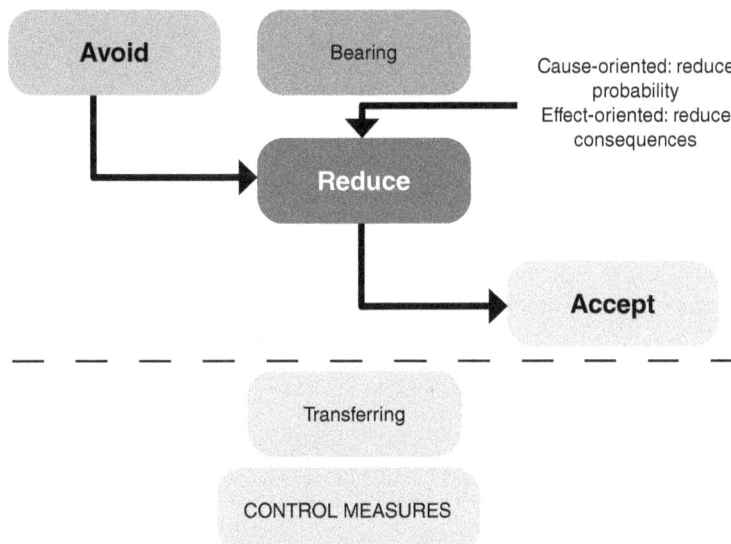

Figure 9.7: Techniques to control construction H&S risk

The construction H&S risk will be borne if it is avoided, reduced or accepted. If the decision is made not to bear the risk, then it needs to be transferred to another party or agency which could be in the form of insurance or contracting out.

Within these techniques the following options are available:
- Physical
- Procedural
- Education and training
- Legal.

Any, or a combination of these options, may be used to achieve the technique decided on. To explain how they may be used in practice, the following examples may be helpful:

PRACTICAL EXAMPLES OF RISK CONTROL

Using the physical options

Avoidance: Not to erect a structure or facility within the 100-year flood line of a river

Elimination: Having stand-by equipment on hand to eliminate any production loss due to the breakdown of a vital piece of construction equipment

Risk reduction: (Frequency). A security fence to reduce the incidence of illegal entry onto the construction site

Risk reduction: (Severity). A high-pressure release valve to prevent a vessel from over-pressurising, limiting damage to the vessel such as an oxygen tank

Risk transfer: Removal of hazardous substances by a specialist contractor.

Using the procedural options
- Avoidance: Prohibiting the use of certain chemicals on a construction project
- Elimination: Safe work procedures, eliminating unsafe acts
- Risk reduction: (Frequency). Buying procedure to ensure compatibility
- Risk reduction: (Severity). Evacuation procedure
- Risk transfer: Procedure to have hazardous waste removed by a contractor.

The option of education and training
- Risk avoidance: On-the-job training
- Risk elimination: First aid training to eliminate the severity of injuries by applying proper treatment
- Risk reduction: (Frequency). Fire prevention programmes to reduce frequencies of fire
- Risk reduction: (Severity). Fire drill to reduce the severity of a fire
- Risk transfer: Supervision.

Using the legal option
- Risk avoidance: Parking on the site on own risk notice
- Risk elimination: Following the manufacturers' requirements
- Risk reduction: Complying with all legal requirements
- Risk transfer: Insurance contract, disclaimer notice or inspection by an accredited authority.

9.12 RISK EXPOSURE PREVENTION PRINCIPLES

The following principles will assist in preventing exposure to risks and experiencing the consequences of such exposure:
- Avoid the risks completely.
- Evaluate the risks that cannot be avoided and develop mitigating interventions.
- Combat the risks at source by considering alternatives and substitutes.
- Adapt to technical progress by using safer technologies, materials and equipment.
- Replace dangerous with non-dangerous using process of substitution.
- Adapt work to worker – design of workplace, choice of equipment and methods.
- Develop a coherent prevention policy – technology, work organisation, working conditions and social relationships, for example:
    - giving collective protective measures priority over individual protective measures such as a scaffold instead of a ladder
    - giving appropriate instructions to workers in the form of safe work procedures, toolbox talks and pre-task briefings.

9.13 RISK CONTROL STANDARDS

Risk control measures should be applied in the form of standards that are to be met. Standards are normally grouped according to risk control disciplines, for example, fire standards, security standards, and occupational health and safety standards.

Standards are based on the following criteria:
- Legal requirements – both common law and statutory requirements
- South African Bureau of Standards (SABS) specifications and Codes of Practice (CoPs)
- Accepted risk management practices
- The collective advice of risk control consultants, brokers and insurers
- The experience and knowledge in the organisation itself.

1. Given that there are two types of person in terms of how they view risk, explain which type you are.
2. Why is it important to conduct a full HIRA, which includes risk prioritisation and quantification?
3. What is the most important aspect of a risk assessment process?
4. Discuss the various options when deciding to transfer a risk.
5. Why would you consider introducing a risk management programme in your organisation?
6. Complete the matrix below using other examples for controlling risks in your organisation.

Techniques or options	Avoidance	Elimination	Reduction	Transfer
Physical				
Procedural				
Education and Training				
Legal				

REFERENCES

European Union. 2014. *The Business Case for Safety and Health at Work: Cost-benefit Analyses of Interventions in Small and Medium-sized Enterprises.* [online] Osha.europa.eu. Available at: https://osha.europa.eu/en/publications/business-case-safety-and-health-cost-benefit-analyses-interventions-small-and-medium

Gauthier, F, Chinniah, Y, Burlet-Vienney, D, Aucourt, B & Larouche, S. 2018. Risk assessment in safety of machinery: Impact of construction flaws in risk estimation parameters. *Safety Science*: 421–433

Haupt, T. 2019. Notes of CEng8042 Risk Management course delivered at Bahir Dar University, Ethiopia, April 2019

Haupt, TC. 2021. *Management of Safety, Health and Environment in South Africa: A Handbook.* Newcastle upon Tyne: Cambridge Scholar Publishing

Health and Safety Executive (HSE). 2014. *Risk Management: Cost Benefit Analysis (CBA) Checklist.* [online] Hse.gov.uk. Available at: https://www.hse.gov.uk/risk/theory/alarpcheck.htm

Hughes, B & Marullo, J. 2009. *Understanding Risk Control — Occupational Health & Safety.* [online] Occupational Health & Safety. Available at: https://ohsonline.com/Articles/2009/10/02/Understanding-Risk-Control.aspx

International Organization for Standardization. 2018. ISO 31000:2018(en) Risk Management — Guidelines. [online] Iso.org. Available at: https://www.iso.org/obp/ui/#iso:std:iso:31000:ed-2:v1:en

Nunes, I. 2016. *OSHwiki*. [online] Oshwiki.eu. Available at: https://oshwiki.eu/wiki/Occupational_safety_and_health_risk_assessment_methodologies

Occupational Safety and Health Administration (OSHA). 2020. *Risk Management Principles and Application*. [online] Oshatrain.org. Available at: https://www.oshatrain.org/notes/4bnotes02.html

Occupational Safety and Health Administration (OSHA). nd. *Hazard Identification and Assessment | Occupational Safety and Health Administration*. [online] Osha.gov. Available at: https://www.osha.gov/shpguidelines/hazard-Identification.html

Smith, N, Merna, T & Jobling, P. 2006. *Managing Risk in Construction Projects*. Oxford: Blackwell Publishing

USEFUL WEBSITES WITH HEALTH AND SAFETY INFORMATION

- Access Point: https://accesspoint.org.uk/technical-loading-bay-design/
- British Occupational Hygiene Society: www.bohs.org
- British Psychological Society: www.bps.org.uk
- British Safety Council: www.britsafe.org
- British Safety Industry Federation: www.bsif.co.uk
- BSI Group: www.bsigroup.com
- Chartered Institute of Environmental Health: www.cieh.org
- Civil Engineers Forum: https://civilengineersforum.com/wp-content/uploads/2014/12/scaffolding-safety.png
- Construction Safety: https://www.constructionsafety.co.za/special-areas/heights-scaffolding/
- Crane Hunter: https://www.cranehunter.com/how-to-read-crane-load-chart
- Department for Environment, Food and Rural Affairs: www.defra.gov.uk
- Department for Health: www.dh.gov.uk
- Department for Transport: https://www.gov.uk/government/organisations/department-for-transport
- Department for Work and Pensions: www.dwp.gov.uk
- European Agency for Health and Safety at Work: www.osha.europa.eu
- European Commission: www.europa.eu
- Faculty of Occupational Medicine: www.fom.ac.uk
- Hazards Campaign: www.hazardscampaign.org.uk
- Hazards Forum: www.hazardsforum.org.uk
- Health and Safety Executive: www.hse.gov.uk
- India Mart: https://www.indiamart.com/proddetail/passenger-cum-material-hoist-21246971848.html
- Innovative Engineering: https://innovativeengineering.com/safety/
- Institute of Ergonomics and Human Factors: www.ergonomics.org.uk
- Institute of Occupational Medicine: www.iom-world.org
- Institution of Occupational Safety and Health: www.iosh.co.uk
- International Institute of Risk and Safety Management: www.iirsm.org
- International Labour Office: www.ilo.org
- National Archives: www.nationalarchives.gov.uk
- National Institute for Communicable Diseases. Frequency of cleaning and disinfection for items in health care workplace: https://ncid.ac.za/wp-content/uploads/2020/05/ipc-guidelines-covid-19-version-2-21-may-2020.pdf
- National Examination Board in Occupational Safety and Health: www.nebosh.org.uk
- Office of the Rail Regulator: www.rail-reg.gov.uk

- Royal College of Nursing: www.rcn.org.uk
- Royal Environmental Health Institute of Scotland: www.rehis.com
- Royal Society for the Prevention of Accidents: www.rospa.com
- SA Ladder: https://www.saladder.co.za/swivel-safety-feet/
- Safework Australia. Frequency of cleaning and disinfection for items in non-health care workplace: https://www.safeworkaustralia.gov.au/doc/how-clean-and-disinfect-your-workplace-covid-19
- Safe Site: https://safesitehq.com/osha-ladder-safety/
- Safety Aspects LLC: https://safetyaspectsllc.com/onsite-safety-training/scaffolding-fall-protection/
- Scaffold Pole: https://scaffoldpole.com/putlog/
- Trades Union Congress: www.tuc.org.uk
- Velvet Cushion: https://www.velvetcushion.com/other/overhead-crane-safety-101
- WHO. Frequency of cleaning and disinfection for items in health care workplace: https://www.who.int/publications-detail/cleaning-and-disinfection-of-environmental-surfaces-in the-context-of-covid-19
- Worksafe New Zealand: https://worksafe.govt.nz/

INDEX

Note: A reference to a page in italics is to a Figure or Table.